D0772940

Multiprocessor Systems on Chip

Torsten Kempf · Gerd Ascheid · Rainer Leupers

Multiprocessor Systems on Chip

Design Space Exploration

Torsten Kempf
RWTH Aachen University
Institute for Integrated Signal Processing Systems (ISS)
Sommerfeldstr. 24
52074 Aachen
Germany
torsten.kempf@iss.rwth-aachen.de

Rainer Leupers
RWTH Aachen University
Software for Systems on Silicon
Templergraben 55
52056 Aachen
Germany
leupers@iss.rwth-aachen.de

Gerd Ascheid
RWTH Aachen University
Institute for Integrated Signal Processing Systems (ISS)
Walter-Schottky-Haus
Room 24 A 207
Sommerfeldstr. 24
52074 Aachen
Germany
ascheid@iss.rwth-aachen.de

ISBN 978-1-4419-8152-3 e-ISBN 978-1-4419-8153-0
DOI 10.1007/978-1-4419-8153-0
Springer New York Dordrecht Heidelberg London

Library of Congress Control Number: 2011921340

Printed on acid-free paper

Springer is part of Springer Science+Business Media (www.springer.com)

Dedicated to Meike and Flora,
to my brother Tibor and
to my parents Brigitte and Wolfgang.

Preface

This book highlights the research conducted in the area of Multi-Processor System-on-Chip design for more than five years. The work documented within was carried out during my time at the Institute of Integrated Signal Processing Systems (ISS) at the RWTH Aachen University.

More than putting forth a brilliant idea, the conducted work reflects a careful evolution of design methodologies and associated tooling. The original motivation dates back to the GRACE++ methodology. This early attempt of system level modeling with SystemC targeted the efficient and convenient exploration of complex architectures, with particular focus on communication architectures. The tight links to industry partners and the ongoing development turned this technology into a commercialized tool called Architects View Framework.

At the time I joined the ISS as a researcher, plenty of experience had been gained in modeling System-on-Chip platforms. By the investigation of several industrial platforms, we soon discovered that the detailed modeling of processing elements limited the capabilities of design space exploration. Accordingly, we extended the methodology to a more abstract modeling of processing elements and, furthermore, broadened it to capture the challenges of temporal and spatial task mapping. With the help of many partners from different research cooperations, we have evolved the methodology and were lucky to be able to validate our approach with relevant design problems. Finally, this innovative technology was brought to the market and became commercially available in 2009.

All the design issues to be found in the development of MPSoC platforms cannot be mastered by a single person. Therefore, I am grateful for the strong support of researchers with whom I had the pleasure to work.

First of all, I would like to thank my supervisor and Prof. Gerd Ascheid who is the co-author of this book. Apart from his valuable feedback and deep interest in my work, I enjoyed the creative working atmosphere of independent research while being guided by inspiring discussions. In the same way, I would like to thank my co-examiner and co-author Prof. Rainer Leupers for his support and valuable feedback.

As mentioned before, my work is based on the Architects View Framework developed by Tim Kogel. Not only for supervising my master's thesis, but also for the joined research projects, I would like to convey my gratitude to Tim.

In addition, I would like to thank my former colleague and office-mate Andreas Wieferink who recruited me to the ISS when I was an undergraduate student. He was always helpful in solving critical debugging issues.

I am grateful to all my colleagues at ISS, who supported me in my research work. Among them I would like give my special thanks to Filippo Borlenghi, Jeronimo Castrillon, Anupam Chattopadhyay, Meik Dörpinghaus, Felix Engel, Lei Gao, Niels Hadaschik, Manuel Hohenauer, David Kammler, Kingshuk Karuri, Stefan Kraemer, Hanno Scharwächter, Stefan Schürmans, Martin Senst, Martin Witte and Diandian Zhang.

When performing research in the area of EDA tools, I personally consider tight interaction with semiconductor and EDA companies as essential to address the key design issues. Luckily, at ISS I had the unique opportunity to meet many helpful professionals over the years, which gave constant guidance and valuable feedback. My special thanks are due to Xavier Buisson, Andreas Hoffmann, Karl Van Rompaey, Bart Vanthournout from CoWare/Synopsys, and to all the professionals we met during the roadshow of the Virtual Processing Unit (VPU).

Converting my ideas into usable tools would have not been possible without the help of my postgraduate students. I would like to thank all of them for their efforts and hard work. Among them, I would like to give special thanks to Jens Reinecke and Stefan Wallentowitz. Furthermore, I would like to thank Filippo Borlenghi, Jeronimo Castrillon, and James Wood for reviewing this book.

I would like to thank my parents for all the constant love and support. I also thank my brother for his support and advice. My very special thanks go to Meike and my daughter Flora for their support, love, and patience.

July 2010 Torsten Kempf

Contents

List of Figures

List of Tables

Chapter 1
Introduction

Over the past 20 years, advances in digital wireless communication technologies have modified everyone's day-to-day life. Predominantly utilized by business customers, the switch from analog to digital wireless communication networks has made them affordable and widely accepted within the consumer market. This trend is clearly reflected by the increase in the number of global cellular subscriptions over the last decade. Figure 1.1 illustrates the impressive growth from $\sim$200 million to over 3,000 million subscriptions listed between the years 1997 and 2008 [1].

Parallel to the achievements in wireless communication, user devices have evolved at an incredible pace over the last years. The technology advances in the semiconductor industry have led to supercomputers in the form factor of a mobile terminal. Accordingly, latest-generation smartphones are no longer limited solely to pure voice communication, but support a wide range of applications from the domains of multimedia, entertainment, and infotainment. In turn, these applications have had a particularly strong impact on connectivity requirements, resulting in the need for the latest smartphones to support multiple wireless communication standards.

These requirements have created one of the most challenging assignments in engineering today. Looking purely at the necessary computational performance shows an approximate demand of 10–80 GOPS peak performance [2] for the execution of today's communication standards. In addition, upcoming standards will further increase the demands, e.g., the upcoming Long Term Evolution (LTE) standard extension. The demand to support the mobility of battery powered devices, makes high energy efficiency one of the key elements for business success within the anticipated market. This demand together with the requirements of low cost, short time-to-market, and the extremely short lifecycles put particular pressure on system architects when designing such terminals.

Today we are witnessing a complete change in the design philosophy of wireless communication devices. In the past, the main answer to the increasing requirements came from the semiconductor technology scaling the manufacturing process, leading to higher performance and energy efficiency gains. These gains were predicted by Moore's Law [3] and Dennard's Scaling Rules [4]. Unfortunately, having reached process manufacturing sizes below 65 nm, further downscaling is becoming more and more challenging, and pessimistic voices predict the end of Moore's Law.

T. Kempf et al., *Multiprocessor Systems on Chip: Design Space Exploration*,
DOI 10.1007/978-1-4419-8153-0_1,

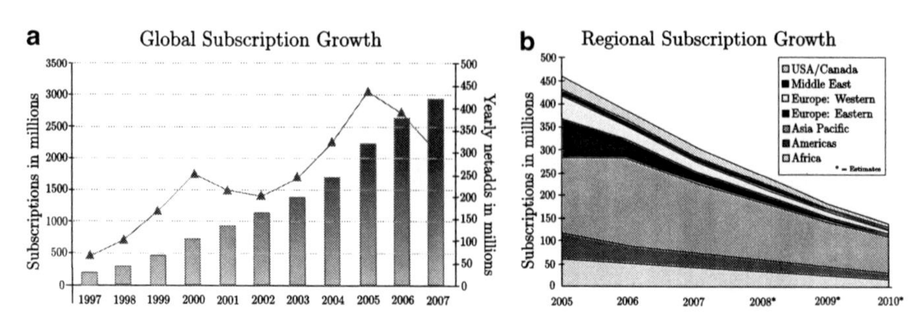

Fig. 1.1 Wireless communication subscriptions (Source: Informa Telecoms & Media [1]). (**a**) Global Subscription Growth and Netadds. (**b**) Regional Subscription Growth

Whether true or not, a more severe design issue has arisen, commonly referred to as the *crisis of complexity* [5]. It is the limitation to fully exploit the advantages provided by process technology due to the lack of efficient design methodologies and tools.

More than ever before, system architects are being required to apply new and innovative designs to increase computational performance and to keep pace with consumer expectations. In a nutshell, the strong computational requirements are forcing system architects to incorporate parallel processing as it offers the capability of sharing the computation among the different resources. Besides this, the contradictory requirements of performance, energy efficiency, and flexibility can only be resolved by programmable processor cores. Starting from the well-known concepts of general purpose computing (GPPs) and digital signal processing (DSPs), the urgent demand for high energy efficiency has led to extensive research on processor cores. One result of this research are application-specific instruction-set processors (ASIPs), which are optimized for a specific application. Furthermore, reconfigurable ASIPs (rASIPs), including postfabrication reconfigurability, have been envisioned and first prototypes are available.

With processor cores being the heart of every wireless communication platform, heterogeneous Multi-Processor Systems-on-Chip (MPSoC) are widely considered to be the optimal choice for implementation. Experiments have shown that, when designed carefully, MPSoCs have the potential to achieve the best trade-off among computational performance, energy efficiency, and flexibility. Unfortunately, system architects are experiencing new and still unsolved challenges during design of such systems. These challenges cover engineering issues ranging from macro- to microscopic aspects in hardware and software development. In addition, earlier design strategies focussing on single components need to be reconsidered, because nowadays only a joint analysis enables statements about the platform capabilities. These issues and challenges have created the research field of ESL design.

Evolving from the fundamental ideas of HW/SW codesign and later system level design, ESL design covers a large set of methodologies and tools surrounding MPSoC design in general. The centerpiece of nearly all ESL design techniques is a virtual platform that serves as an executable specification to evaluate particular design objectives. Virtual platform techniques have achieved a major break-through in the fields of software development and debugging, as well as platform analysis, optimization, and verification. These virtual platforms replace costly hardware prototypes and have the potential to significantly simplify and speed-up the design process. However, virtual platforms operating on instruction-set level can hardly be used directly at the start of the design cycle, when typically neither the hardware architecture nor the compiler tool chain and/or the software implementation are fixed. Therefore, innovative design methodologies to carry out early design space exploration are essential, as last minute design changes tend to be extremely costly and induce high risks of wasting development effort. Accordingly, these methodologies have to support system architects in identifying the optimal or suboptimal design options right from the outset. Moreover, for wide acceptance and practical use, a clear link to existing technologies is mandatory.

To address the design issues of future multi- and many-processor core architectures, with particular attention to platforms in the domain of wireless communication, this book outlines a unique early design space exploration framework. Its major contribution is a joint environment that covers several abstraction layers for the purpose of the exploration and evaluation of heterogeneous MPSoC platforms. The framework introduces the following main concepts and techniques and its overall structure is comprehensively described in Fig. 1.2.

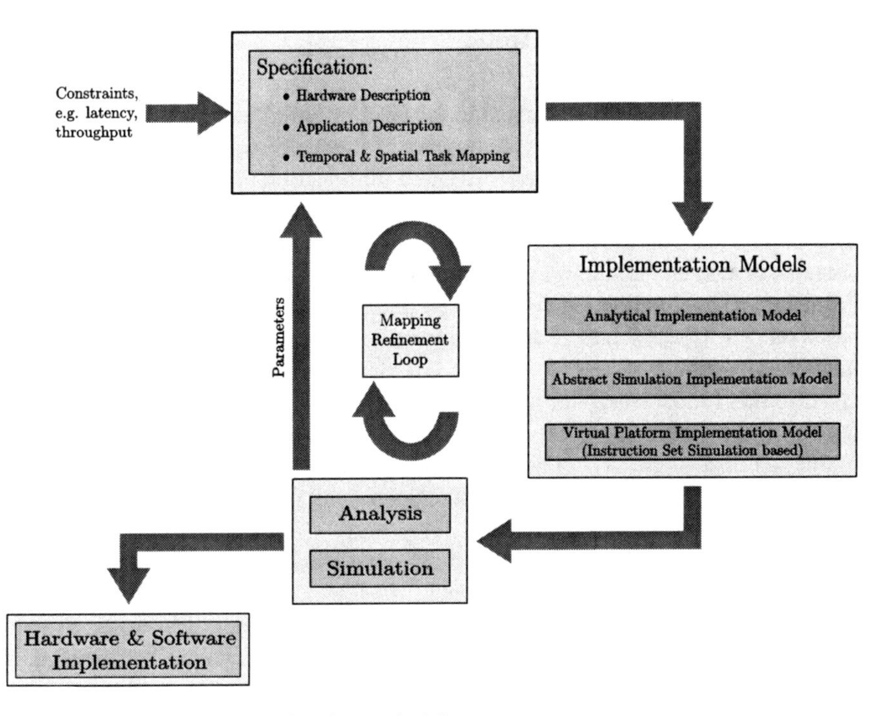

Fig. 1.2 Early design space exploration methodology

- An *analytical implementation model* is built on the fundamentals of statistical processes and graph theory. This model targets early design stages, when the hardware architecture is undefined or only a few components are available. The key idea is to formally describe the anticipated hardware platform, the application specification and the temporal and spatial task mapping at a high abstraction level. Based on this solid foundation, a mathematical analysis allows the computation of the performance characteristics and helps to identify whether the system complies with the necessary constraints and also to highlight potential design difficulties.
- The second major implementation model is based on an *abstract simulation model*. The key principle is an annotation of the execution characteristics supporting the evaluation of arbitrary aspects without a detailed and time-consuming implementation. This paradigm has culminated in the Virtual Processing Unit (VPU) and several extensions for practical use and the investigation of common hardware features.
- Acceptance and usability not only require sophisticated implementation models but also an *effective design process* with the possibility of a smooth transition between the abstraction layers. In strict adherence to this paradigm, the proposed framework provides techniques to (semi-)automatically close the design gaps between the abstraction levels.

1.1 Organization of the Book

Various research activities dedicated to the field of multi- and many-core architectures have generated a considerable number of methodologies and techniques. For this reason, this book gives a rather detailed introduction into the overall ESL domain. Chapter 2 discusses the general application and architectural trends and also their implications on the design methodology. This includes applications from the domains of wireless communication, multimedia, and other general purpose ones. From the architectural perspective, utilized IP components are separately introduced based on their type, such as processing elements, communication architectures, and memories.

After this fundamental introduction, Chap. 3 identifies and highlights the foundation of any design space exploration. Central aspects that are discussed are the evaluation of a single design point and the strategy to navigate the design space. The chapter concludes with the identification of the requirements for an efficient design process and framework.

As the proposed framework is definitely not a single entity within the complex research space, Chap. 4 depicts the related work which can be found in academia and industry. The chapter is divided into two aspects, namely Electronic System Level (ESL) design and early design space exploration, covering both analytical and simulation-based approaches.

The subsequent chapters introduce the proposed methodology and framework for early design space exploration. Chapter 5 highlights the overall principle and structure of the methodology which follows the paradigm of abstraction. The analytical implementation model is situated at the highest abstraction level, whereas the abstract simulation model bridges the design discontinuity to the well-known ESL design at the level of instruction set simulation. As a consequence, a continuous design process from a high- to low-level of abstraction is inherently ensured.

Chapter 6 discusses the analytical implementation model. Within the discussion, the problem of design space exploration and analysis is defined as a mathematical problem. Finally, the chapter concludes with the link to the abstract simulation-based environment.

The abstract simulation model is discussed from a practical point of view in Chap. 7, which highlights its practical usage and introduces the underlying concept, as well as the provided features. Subsequently, the refinement from the abstract to the instruction set simulation model is presented.

In Chap. 8 the usefulness and accuracy of the proposed framework and underlying concept are proved by a case study from the domain of wireless communication. This case study covers two main aspects. The first part captures the accuracy that can be achieved for various design decisions and the different modeling techniques, whereas the second part highlights the practical use based on a complex, yet typical design process.

Finally, the book concludes with a summary and an outlook on further research in the field of design space exploration.

Chapter 2
Systems for Wireless Communication

The advent of second generation (2G), digital mobile communication networks for the mass markets had a significant impact on the use of mobile communication in the 1990s. Previously, the usage of mobile communication had been limited to business customers because of the high costs, whereas second (2G) and following (3G, LTE) wireless communication generations have been affordable for the masses. With the change of customers, the usage of mobile communication has broadened from pure mobile voice communication to infotainment and entertainment. This requires mobile handsets to support, in addition to the key components of voice and data communication, applications, like multimedia ones. The different structure and demands of these applications require different kinds of wireless communication protocols and standards which, in turn, has led to the incorporation of a hardware subsystem for each standard. This solution promises short-term success, however in the long term this principle is not expected to scale with a large number of supported communication standards. Finally, this has led to the vision of a Software Defined Radio (SDR) [6] which implements these standards in software to allow an easy upgrade and extension of the set of supported standards. It is commonly agreed that heterogeneous Multiprocessor System-on-Chip (MPSoCs) [7] are the best choice for the underlying platform to cope with the challenging demands of computational performance, energy efficiency, and flexibility, especially for wireless communication devices like SDRs.

This chapter first examines the applications executed on cellphones and smartphones separately for the three domains of wireless communication, multimedia, and general purpose. Based on them, the impact and constraints for the design methodology for wireless communication platforms are derived. The second part of the chapter discusses the underlying hardware platforms and components. Additionally, the specific influence of the platform and components on the design process is highlighted.

T. Kempf et al., *Multiprocessor Systems on Chip: Design Space Exploration*,
DOI 10.1007/978-1-4419-8153-0_2, © Springer Science+Business Media, LLC 2011

2.1 Applications for Mobile Devices

Applications for mobile devices differ significantly in their characteristics according to their domain. Therefore, they are discussed separately. Applications for wireless communications, with particular focus on physical-layer processing, are treated in greatest detail as the case study discussed in Chap. 8 addresses this domain.

2.1.1 *Wireless Communication Domain*

Within this area, targeted applications comprise all kinds of standards and protocols for voice and data communication. To achieve highest interoperability these are typically standardized by organizations like ITU [8], ETSI [9], and IEEE [10]. In addition, the application structure is defined according to the International Standard Organization Open Systems Interconnection Basic Reference Model (ISO/OSI Reference Model) [11] to simplify the design of wireless communication standards. However, modern standard implementations are not too strict about dividing the different layers, so that applied cross-layer optimizations soften the borders between adjacent layers.

A large variety of wireless communication standards have emerged, each addressing a particular range of user-level applications. The traditional classification of standards differentiates among Wireless Personal Area Networks, Wireless Local Area Networks (WLAN), Wireless Metropolitan Area Networks (WMAN), and Wireless Wide Area Networks. Figure 2.1 illustrates these four classes including examples and use-cases. Additionally, localization services like the Global Positioning System (GPS) are considered as a part of wireless communication systems.

The multimedia and wireless communication domains are converging initiated by technology advances, e.g., high performance mobile processor cores, as well as high-resolution displays and touchscreens for mobile devices. The result of this convergence is a class of *smartphones* that combine the functionalities of

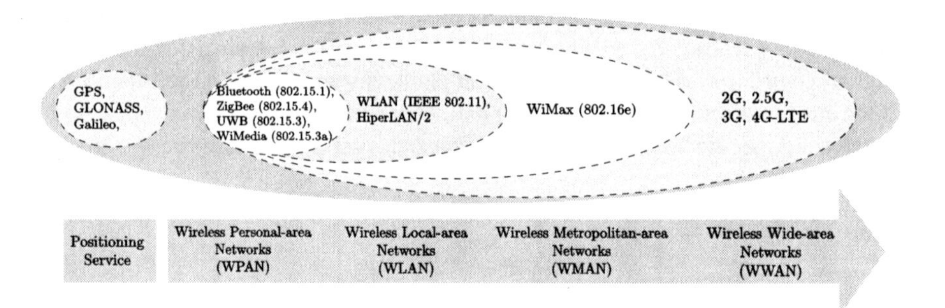

Fig. 2.1 Wireless communication networks

mobile phones and personal computers (PC) into a single mobile device. These devices support a wide range of different applications, each having individual connectivity demands. This requires the support of different wireless communication standards, e.g., Bluetooth [12] for wireless headsets, WLAN [13] for internet access, and 2G and 3G network connection for voice and data communication. Past and present designs cope with this challenge by incorporating one subsystem for each supported standard. For example, Apple's 3G iPhone [14] includes five subsystems for GSM/GPRS/EDGE (2G), WCDMA/HSDPA (3G), GPS, WLAN, and Bluetooth [15]. The addition of further subsystems to support additional wireless communication standards is not expected to scale in future. To cope with this issue, industry and research have opted for SDR [6], where different wireless communication standards are implemented in software allowing the reuse of hardware components. A case study [16] carried out by Infineon expects an SDR to outperform the traditional solution in terms of area and costs at five implemented standards. However, implementing even a single wireless communication standard is already a complex task, therefore the design of a complete SDR becomes quite challenging.

The development of wireless communication standards is dominated by the physical-layer processing, i.e., the lowest layer in the ISO/OSI reference model. This layer has a high computational demand (10–80 GOPS) and (mostly) hard real-time constraints have to be fulfilled. From the application perspective, the most severe constraints are *latency* and *throughput*. Failing to comply with these constraints will most likely lead to business failure.

The key elements of physical-layer processing are digital signal processing algorithms. These algorithms are typically characterized by a computationally intensive data-plane processing at high data rates predominantly controlled through parametrization. These data flow dominated applications are rather well structured in terms of *task graphs* or block processing, allowing the utilization of static schedulers and making task-level parallelism rather clear. The known task structure and data communication between different tasks can be easily captured in task graphs, e.g., Kahn Process Networks (KPN) [19] or Synchronous Data Flow (SDF) [20] task graphs. For specific task graphs, especially for the latter mentioned SDFs, a static schedule can be derived prior to run-time. Thus, deterministic behavior is ensured and no dynamic overhead occurs. Figure 2.2 exemplifies a task graph structure of a WLAN 802.11a receiver [17] implementation.

2.1.2 Multimedia Applications

The domain of multimedia covers a wide range of applications like audio, image, and video processing along with 2D and 3D graphic applications. Similar to the wireless communication domain, many standards coexist in the field, each having a particular optimization criterion like data compression or high quality.

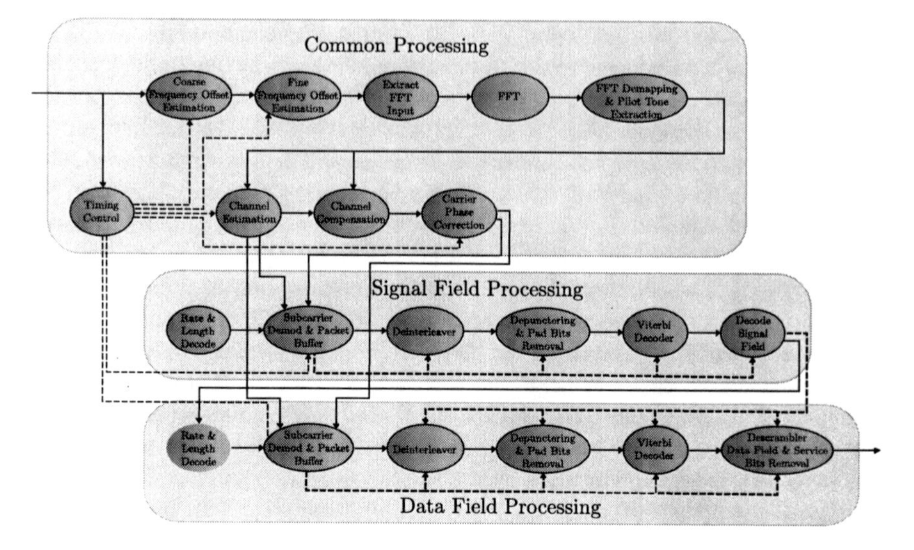

Fig. 2.2 Wireless communication task graph example: WLAN 802.11a receiver [17, 18]

Table 2.1 Computational and communication requirements of multimedia applications [21]

Typical configuration		On-chip communication requirements		Computational requirements in operations per second (GOPS)
Resolution	Frequency (Hz)	Pixels per second (Mpxl/s)	Bytes per second (MBps)	
720 × 480	60	20.7	82.8	31.1
1,280 × 768	60	59.0	236.0	123.9
1,920 × 1,080	30	62.2	248.8	186.6
1,920 × 1,080	60	124.4	497.6	373.2
1,920 × 1,200	60	138.2	552.8	414.7

Multimedia applications are characterized by a *high computational* demand along with *high communication* requirements between processor and memory. The key application in this domain is video processing with requirements including stringent real-time constraints as exemplified in Table 2.1. This especially applies to high-quality video applications, which include high throughput demands with challenging latency requirements.

These challenging demands of the latest multimedia standards along with the need for energy efficiency, in particular for battery-powered mobile devices, has led to highly specialized hardware accelerators and graphic-processing units [22]. These hardware components are optimized for performing typical 2D and 3D graphic processing operations, e.g., texture mapping and rendering especially vertex, geometry, and pixel shader calculations. The key characteristic of these algorithms is the *massive parallelism* of vector and matrix operations. Similar to ASIPs, these hardware architectures are especially tailored for the needs of multimedia applications. The design principle is to restrict flexibility to a minimum to achieve highest

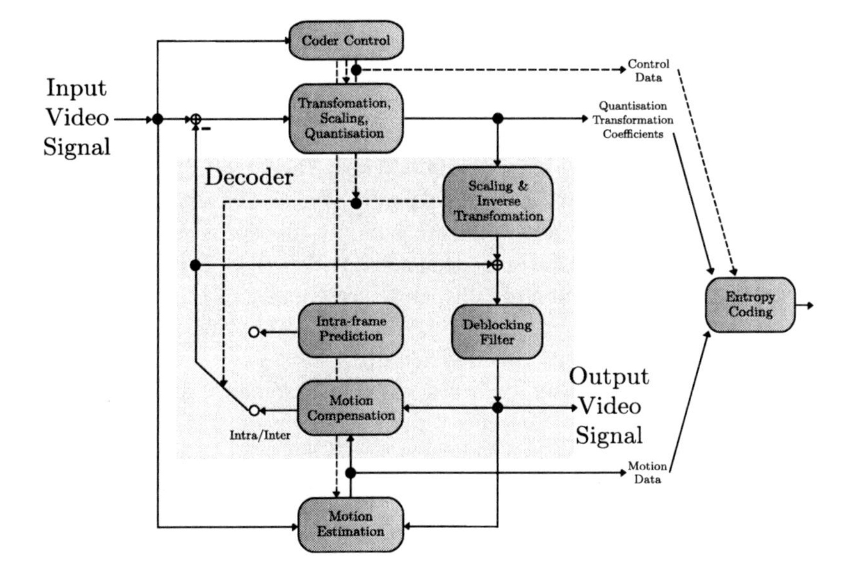

Fig. 2.3 Multimedia example H.264 task graph [23]

performance and energy efficiency. Despite the immense performance provided by such architectures, software development for them is extremely challenging. In addition, identifying the inherent parallelism within a particular algorithm is key and is mostly carried out by application experts and manual interaction.

Multimedia standards are mostly defined as task graphs (Fig. 2.3) like applications from the domain of wireless communication. The included control flow, e.g., the control overhead in H.264 decoding, leads to severe challenges in memory optimizations and data communication. Therefore, implementations commonly require special treatments to optimize the data communication.

2.1.3 General Purpose and Other Applications

Various kinds of applications are categorized under the term of general-purpose applications. Typical examples are text processing and web-browsing applications. Traditionally developed for personal computers (PCs), these are becoming increasingly popular even on mobile devices like smartphones. These applications are software-centric and make heavy use of operating systems (OSs), middleware layers, and other forms of hardware abstraction layers (HAL) such as hardware dependent software (HdS). Contrary to performance-critical parts, like physical-layer processing in the domain of wireless communication, they have a less dominant data plane processing and their computational complexity is rather low. On the other hand, control plane processing is much more severe because applications have to react on *nondeterministic* user interactions.

This leads to a complex control flow execution, which requires techniques for efficient execution, such as efficient implementations of jump and branch instructions as well as function and procedure handling. To accelerate them, well-known personal computer techniques, such as like branch prediction and superscalar architectures [24], are being increasingly adopted. This architecture trend is constantly narrowing the gap between general purpose processors within the embedded system and the personal computer market. Naturally, this opens new market opportunities for IP vendors from the embedded domain like ARM, MIPS, and Tensilica while moving into the direction of GPC. However, companies originating from the domain of personal computing, e.g., Intel, AMD, and VIA, have announced or are already are moving toward embedded systems [25].

In contrast to the previously discussed application domains, several description and development techniques for general purpose applications exist. However, the most common method is the classical textual design based on a high-level programming language based on C/C++ or Java. Other approaches like component-based software design [26] or the unified modeling language (UML) [27] provide graphical design entries for improved implementation efficiency.

2.1.4 Application Impact on Design Methodology

The rapidly increasing performance demands and limited available energy of battery powered devices, gives rise to an increasing energy-performance gap. Additionally, the need to jointly support various applications and their requirements is having a significant impact on the design methodology. *General purpose applications* demand flexible architectures to support a wide range of applications. Characterized by a dominant control path, software development relies on high-level programming languages along with operating systems (OSs), middlewares, and libraries. In contrast, applications from the domain of *wireless communication* and *multimedia* are implemented by highly specialized architectures and low-level software development. Applications of these domains are characterized by high computational demands in the data plane processing with relatively low control overhead.

In general, these various application requirements have significant impact on the design methodology of the two major components *software* and *hardware*. From the hardware perspective, the complexity and computational demand of modern communication standards requires rapidly increasing performance while preserving energy efficiency for future wireless communication devices. As the current technology scaling cannot cope with these requirements by itself, new approaches have to be considered [5]. An obvious solution is to apply parallelism, in terms of processing the application on multiple processing elements in parallel. In addition, the contradictory requirements of high computational power and energy efficiency require highly specialized hardware architectures. This has led to the common agreement that *heterogeneous MPSoC* platforms are the best candidate for such devices [28].

Unfortunately, the selection of heterogeneous MPSoC platforms has a significant impact and induces design challenges like:

- Partitioning of tasks to optimally exploit the inherent parallelism within a given application.
- This partitioning is tightly linked to the selection of the type and number of hardware components, which is a key question for assembling the hardware architecture.
- Performance evaluation can no longer be performed on the basis of a single isolated component. Instead, the interacting behavior of all system components requires a system-wide performance evaluation.
- New programming techniques and models need to be considered since, due to the heterogeneous nature of the platform a simple adaptation of known multiprocessor programming is not feasible.

In addition, the first and most important design objective is to achieve the performance requirements, mostly given in regard to *latency* and *throughput* constraints. These requirements, particularly when implementing the physical layer of a wireless communication standard, are characterized by stringent (hard) real-time constraints that have to be fulfilled. Otherwise, devices will most likely fail standard compliance tests, leading to business failure. Hence, the design methodology must incorporate techniques to efficiently evaluate whether the application-induced constraints are met or not. As late design changes tend to be more costly than early ones, such techniques should be applied as early as possible in the design process.

After discussing the application needs and their coarse-grained impact on the design methodology, the discussion now turns to detailed design aspects and the corresponding influence of each possible hardware component. Along with this, the impact on the design methodology is highlighted.

2.2 Hardware Platforms and Components

New design methodologies offering increased productivity in terms of design efficiency are indispensable for the development of future heterogeneous MPSoCs. For the comparison of different MPSoC platforms the following fundamental objectives and metrics can be defined.

Performance. Probably the most important design objective, the performance, is typically measured in terms of latency and throughput. Especially, meeting the performance constraints induced by applications is highly challenging but necessary for a successfully operating device.

Energy and Power Efficiency. Energy efficiency is one of the most severe design issues and platform differentiators. Especially, for mobile and battery powered devices energy efficiency is essential. Unfortunately, over the last years battery capacity has not been able to cope with the increasing performance demands, leading to a growing performance-energy gap. This requires architectural innovations to increase the energy efficiency needed at present and definitely in the future. The metric *Millions of Instructions Per Second (MIPS) per Watt* typically defines energy efficiency [29]. Although this rather crude definition gives designers a first rough idea, it is unsuitable, as it is the required *energy per task* which matters. In the domain of wireless communication this metric can be expanded to the required *energy per decoded bit* or, within the domain of multimedia, to *energy per pixel*. The power efficiency classifies the power dissipation on the chip which influences the package and the layout of the final chip.

Cost. In general the total costs consist of the design costs and the initial manufacturing costs [30]. Whereas the design costs include the development of both software and hardware, the initial manufacturing costs comprise the mask and wafer costs as well as the initial packaging and testing. The dominating design costs are related to software and hardware development. These are reported for current design technologies (90 nm) to be in the region of 10–100 million USD with an expected increase of 50–100% per shrink in the process generation. Whereas in the past hardware-development costs claimed the major portion, the increasing use of programmable components has led to rapidly increasing software costs [30]. Latest market studies of MPSoC design report them to be at the same level. In addition, chip mask production has become increasingly expensive and is typically in the range of multiple million USD for each mask iteration.

Flexibility. In contrast to the previously discussed objectives and metrics, flexibility cannot be simply given as a single value. Flexibility defines the capability to execute a specific functionality on a particular processing element. This metric is of vital importance especially when designing SDRs [6]. Additionally, flexibility has the advantages of enabling short time-to-market and extending the lifetime by applying software updates and bugfixes. It is closely related to portability, which defines the ease of porting a certain functionality from one platform to another. Portability can be defined as the inverse of the porting effort [31] which, in turn, directly relates to flexibility.

These objectives help to guide system architects in their design decisions to find the optimal design. However, the complexity and short time-to-market along with the discussed requirements put a particular pressure on the development of such MPSoC platforms. Therefore, new design methodologies have to be considered to minimize the required development effort and costs. Here two fundamental design concepts, namely component-based design (CbD) [32] and platform-based design (PbD) [33], have been envisioned and found major acceptance.

MPSoC design: Evolution rather than Revolution. According to the component-based approach, the complete platform is assembled from in-house or external IP components, e.g., processor cores, communication architectures, memories, and many other IP components. The key to the efficient use of this design principle is a unified interface definition to connect arbitrary IP components. These interfaces are mostly bus or Network-on-Chip (NoC) centric, like the interfaces of the

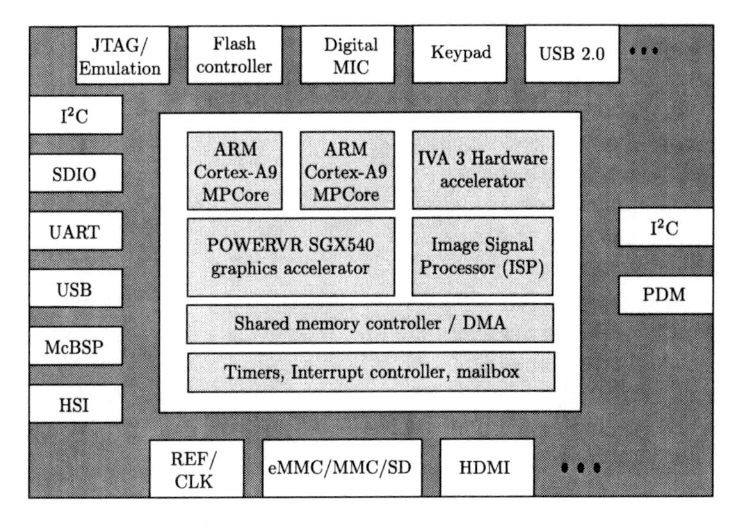

Fig. 2.4 IP block structure of the TI OMAP44x platform [38]

AMBA bus [34] or the IBM CoreConnect [35]. These have been standardized or evolved to a de facto standard by wide utilization. Based on this design methodology a large variety of companies have established a successful IP business, among them processor IP vendors like ARM, MIPS, and Tensilica as well as communication architecture IP providers like Arteris [36] as well as above-mentioned IP vendors like ARM and IBM. An example IP-component structure for TI's OMAP [37] platform is sketched in Fig. 2.4.

This CbD inherently ensures the high reuse of components over different platforms as they are separated by well-defined interfaces. Because of growing complexity, the average number of IP components an MPSoC platform consists of has increased from 25 in 2006 to 28 in 2007 and 33 in 2008 [39]. Further predictions expect an increase over the next years, already reaching 72 IP components in an average platform design by the year 2012.

With the aid of such IP components, PbD has proved to be highly suitable to quickly obtain modified platforms from a base one. The major element is the restriction of the design space by reducing flexibility, which simplifies and shortens the development cycle significantly. This design methodology has been successfully applied to especially address a specific market segment, e.g., the areas of wireless communication and multimedia. Prominent examples are TI's OMAP platforms for wireless communication devices and Philips Nexperia [40] platforms for multimedia applications.

The development of each platform is based on a construction kit. For each market segment a particular set of IP components is selected and connected. For example, the OMAP331 targeting the low-cost segment consists of an ARM926EJ-S processor core with a few surrounding peripheral devices. The high-cost segment is addressed by TI's OMAP3430 [37] platform that includes a more powerful ARM

Cortex-A8 processor, an IVA 2+ graphics accelerator, a POWERVR SGX graphics core [41], a dedicated image signal processor (ISP), and various other peripheral devices.

Apart from the business success of such platforms, this design methodology bears some hidden traps and risks [42, cf. 7.2]. The key risk is that system architects enter the design cycle biased and do not question design decisions related to the preexisting software or hardware IPs. In the end this can lead to false design decisions that decrease performance or increase energy consumption. In contrast, starting designs from scratch without reusing pieces of existing platforms is also no option when considering the tight time-to-market constraints. Therefore, a suitable design methodology requires a mixture of both extremes and demands strong design discipline. Hence, Bailey et al. [42] propose that all design options should be considered when developing a modified platform virtually starting with a blank sheet of paper, but characteristics, prior experiences and reuse of existing IP components can be incorporated to enhance the design process and the final platform.

As a large variety of different components exists, the rest of this section discusses each particular group of components separately and highlights the impact on the design methodology. However, it should be noted that the most essential issue in MPSoC design is the interwoven behavior of all the components and not that of a single isolated component. For example, a high performance processor core cannot fully exploit its capabilities if either the communication architecture or the memory subsystem is too slow to deliver the necessary data to be processed. Such issues cannot be evaluated in an isolated fashion because they only occur when investigating the system-wide performance.

2.2.1 Processing Elements

The class of processing elements ranges from highly flexible general purpose processors (GPPs) to dedicated hardwired accelerators, optimized for a particular function. Lately, the demand for postfabrication flexibility has led system architects to increasingly use flexible and programmable components like general purpose processors, digital signal processors (DSPs), and application-specific instruction-set processors. Consequently, the amount and the importance of software are steadily increasing. Already today software has become one of the most critical pieces in system design [43], consuming a significant amount of the overall budget. With the increasing introduction of heterogeneous MPSoCs, various software design methodologies need to be considered jointly ranging from high-to low-level software constructs.

The class of the processing elements can roughly be classified into the following groups.

- General Purpose Processor (GPP)
- Digital Signal Processor (DSP)
- Application Specific Instruction Set Processor (ASIP)

- Reconfigurable Application Specific Instruction Set Processor (rASIP)
- Field Programmable Gate Array (FPGA)
- Application Specific Integrated Circuit (ASIC)

General Purpose Processors offer high flexibility and are hence utilized for arbitrary applications like control and user-level applications. Commonly, application development is conveniently carried out in high-level programming languages, e.g., C/C++ and Java. Often an operating system (OS) is supported and software development is abstracted by HAL or other middlewares from low-level hardware features. This shields software design from the underlying hardware by means of abstraction, permitting to concentrate on the pure application development.

Digital Signal Processors are especially tailored for the common characteristics and operations of digital signal processing algorithms. These processors exhibit special instructions to efficiently perform operations common to these algorithms, e.g., multiply accumulate, add-compare-select, and Galois field instructions [44]. The latest DSP architectures provide increased parallelism by means of Very Long Instruction Words [45], Single-Instruction Multiple-Data [46], and superscalar [47] hardware features. Because of the high-performance and low energy-consumption demands in the domain of wireless communication, fixed-point DSPs are still the first choice even after the introduction of floating-point DSPs [44].

Application Specific Instruction Set Processors are specially developed for a specific application. In general, the design of an ASIP follows the guideline of minimizing flexibility to maximize energy efficiency, area efficiency, and/or performance. Today, the class of ASIPs covers a wide range of different approaches and architectures. Tensilica's approach [48] enters the design process with the Xtensa processor core as a base architecture and allows further customization of this template with respect to the addressed application. Other approaches support ASIP development based on an Architecture Description Language (ADL), e.g., LISA 2.0 [49] or Expression [50]. These ADL-based approaches do not restrict designers in their decisions to support full architectural design space exploration. Contrary to GPPs, application-specific features cannot be easily addressed by compilers. Therefore, ASIPs typically require low-level software development to exploit the specific features. However, there are promising approaches to generate the software tool-chain including compiler, assembler, and linker for the ASIP [51, 52] with reasonable performance.

Reconfigurable Application Specific Instruction Set Processors extend the concept of ASIPs further by combining the base processor with a reconfigurable fabric based on FPGAs [53]. This combination of a fixed and a reconfigurable hardware architecture promises high performance with increased flexibility to adapt the designed processor to different applications. Compared with the previously discussed ASIPs, the reconfigurable part adds postfabrication flexibility. Already a few architectures [54] and design methodologies [55, 56] exist, highlighting the potential of such architectures. However, this research field is relatively new and is expected to have high potential in the future. Besides the earlier-mentioned issues for ASIPs, additional hardware description language (HDL) programming needs to be included to program the embedded FPGA.

Field Programmable Gate Arrays are reconfigurable processing elements. Based on the capability to reconfigure the functionality after manufacturing, these components provide a particular postfabrication flexibility. The utilization of such devices has a strong impact on the design process, because FPGA devices are traditionally programmed in hardware description languages, e.g., VHDL [57] and Verilog [58]. Therefore, adding an FPGA to a platform changes the design process to a mixed software and hardware development. However, its flexibility compared to ASICs is achieved at the expense of decreased performance and increased energy consumption, but offers the possibility of reprogramming and bugfixing in the field.

Application Specific Integrated Circuits are specially tailored for a given algorithm or application. With the functionality fixed, only minor configuration can be applied after fabrication. Mostly this configuration is limited to the setting of algorithmic parameters, e.g., the filter coefficients of an FIR filter. In contrast to the restricted flexibility, energy efficiency and performance are relatively high. This leads to an integration of such processing elements in the performance-critical parts of a design. The traditional design focuses on known hardware design methodologies like Register Transfer Level (RTL), modeling with logic synthesis on standard-cell libraries, or full-custom design on transistor level.

Summarizing the common use of these processing elements, wireless communication and multimedia algorithms, as proved in the past, can be efficiently implemented on specialized hardware. Dedicated hardwired accelerators (ASICs) are especially tailored for a particular algorithm, whereas DSPs are optimized to the common characteristics of such algorithms, e.g., multiplications, multiply accumulate, and add-compare-select. Application Specific Instruction Set Processors (ASIPs), like those proposed by Wehn et al. [59] or SODA [2], are specialized processor cores which have been specially developed for a particular algorithm or multiple ones. The key principle of ASIPs is to minimize the provided flexibility to increase performance and to minimize overheads in terms of area, power and energy consumption. To incorporate such specialized architectures, software development cannot follow the general-purpose approach, as current high-level language compilers can hardly exploit such features optimally due to their highly irregular structure [31,60]. However, research focuses on this issue and promising approaches exist in literature [51,61–64].

In contrast to specialized processing elements, general purpose applications require a higher degree of flexibility. Hence, GPPs are typically utilized for their execution and the latest techniques and architectures from the personal-computer domain are increasingly being applied to mobile devices. For example, ARM Inc. has just recently announced the ARM Cortex-A8 processor core as their first superscalar processor core. Additionally, multicore processors like the ARM Cortex-A9 can already incorporate up to four cores within a single entity.

So far only processing elements have been considered. However, with the increasing parallelism in future platforms, data exchange between the processing elements is becoming another key issue. In general, to transfer data from one element to another, a communication architecture and storage elements are necessary. Recently with the increasing number of interacting components, the principle of bus-based communication architectures has gradually tended to become the bottleneck of the complete system. Therefore, the latest research in this domain has proposed more complex communication networks subsumed under the term Network-on-Chip (NoC) [65].

2.2.2 Communication Architectures and Memory Subsystems

Despite the vast research and many publications within this domain, a precise definition of NoCs is typically not given [66]. The OCP-IP consortium defines the term NoC in a rather generic fashion as *a communication network that is used on chip* [67]. This generic definition allows further differentiation of NoC architectures under the key aspects of:

- *Switching policy*: circuit-switching and packet-switching.
- *Topology*: point-to-point, bus, hierarchical bus, crossbar, 2D-mesh, 2D-torus, 3D-torus, customized for the addressed application, etc.
- *Routing*: deterministic fixed routing and dynamic adaptive routing.
- *Quality-of-Service*: best effort and guaranteed throughput.
- Testing and fault-tolerance.

When dealing with embedded systems, research about NoCs has to adhere to the special demands of this domain in terms of cost, power and energy efficiency [68]. Similar to the application specific processing elements like DSPs, ASIPs, and rASIPs, customized application-specific NoCs achieve superior performance in terms of latency, throughput, area, power and energy efficiency by restricting the flexibility [67]. However, the highly irregular topologies of these communication architectures increase the effort required for wiring and layout. As this book focuses on early design space exploration of heterogeneous MPSoC platforms, interested readers are here referred to [69] for a detailed discussion of available Network-on-Chip architectures and design methodologies.

The general design approach of CbD, treats a memory subsystem as a single hardware IP component due to the highly regular structure that is attached to a particular communication architecture and used to exchange data. The memory portion in modern MPSoC platforms is tremendous and considered to be in the range of $\sim$60% of the complete area [70]. Therefore, area and energy consumption can be significantly reduced by designing efficient memory architectures. A classical hierarchical memory system as illustrated in Fig. 2.5 attaches the processor core directly to a fast scratchpad memory or cache which is further connected to a larger memory and finally over I/O devices to external memories like hard disks and flashcards.

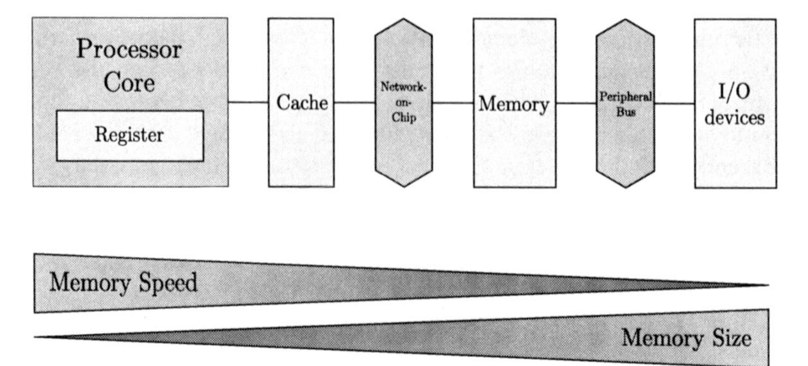

Fig. 2.5 Memory architectures of common processor cores [24]

With respect to early design space exploration, the impact of the memory subsystem has to be evaluated in terms of size and performance. Whereas, the memory size has a significant impact on the required chip area, the performance characteristic depends on the communication between processor and memory. This includes the underlying communication architecture and has to consider the occurring memory access patterns. Therefore, design methodologies have to support designers in performance evaluation of the memory subsystem together with the communication architecture. Key objectives are to minimize the required memory size to reduce area and energy consumption, as well as to increase performance characteristics in terms of latency and throughput.

2.2.3 Hardware Architecture Impact on Design Methodology

The chosen hardware architecture heavily affects whether the application performance constraints are met or not (Sect. 2.1) and how the platform behaves with respect to the formulated objectives (Sect. 2.2). An accurate evaluation of the performance and the objectives can only be done when based on the complete platform and not on the investigation of a single component. However, selecting each processing element can have significant impact on the design methodology.

Looking at the programmable part of the architecture, GPPs execute general purpose based applications that are typically developed in a high-level programming language including the use of operating systems, middlewares, and hardware abstraction levels. Conversely, performance critical applications execute on more application-specific architectures like DSPs, ASIPs, or even weakly programmable devices [71]. In general, such specialized architectures require hand-optimized low-level software implementations. Unfortunately, this influences the design methodology, so that time and cost intensive software development [72] is certainly required to achieve the necessary performance. Therefore, prior to the implementation step, system architects *must* evaluate that the implementation of software and hardware satisfies the addressed performance requirements. Otherwise late design changes, e.g., exchanging the type of a processor core, will result in high time and cost investment as optimizations need to be applied twice [73]. This demands a sophisticated evaluation methodology for the early investigation of different design decisions leading finally to the implementation of both software and hardware. Hence, a joint hardware/software codesign including different software development techniques is indispensable.

In addition to the use of software-centric processing elements, performance-critical parts might require less flexible processing elements due to high computational requirements and/or low flexibility demand. In such cases components like ASICs, FPGAs, but also rASIPs should be considered. These processing elements extend the software-centric design flow to also incorporate traditional hardware-design methodologies. This joint consideration of software and hardware makes a mixed hardware/software codesign essential and tends to be complex.

Besides the impact of the processing elements, the impact of the communication architecture must not be neglected. In the high-performance computing domain more general and regular NoCs are selected, whereas in the area of embedded systems customized NoCs dominate the market. These NoCs are optimized for the particular needs of one or multiple applications and restrict the flexibility by limiting the number of physical links, which in turn restricts the mapping space of the application. Therefore, the development of these application-specific communication architectures requires a deep knowledge of all the other hardware components and the addressed applications. This leads to the necessity for a joint design method, including the investigation of arbitrary communication architectures.

In summary, the challenging performance requirements and objective constraints force system architects to utilize heterogeneous MPSoC platforms, which, in turn, significantly affects the software design methodology for such platforms.

2.3 Summary

So far this chapter has highlighted the impact of applications, hardware and software decisions on the design methodology. Driven by the complexity and requirements of future applications system architects are increasingly using heterogeneous MPSoC platforms. These platforms are commonly assembled from different IP components that can be roughly grouped into the classes of processing elements, communication architectures, and memories. Based on the selected processing element used to execute a particular functionality, hardware/software codesign becomes necessary. This becomes a key challenge, especially when utilizing application-specific and optimized processing elements. In addition, different software-development schemes have to be used jointly to optimally exploit the features of the underlying hardware.

The range of applied software development has to consider all possibilities, starting from low-level Assembly implementation up to high-level programming languages and the use of operating systems, middlewares, and other HAL.

The various challenges in designing such MPSoC platforms demand a structured design methodology. The next chapter discusses the principles of what is traditionally considered *design space exploration*.

Chapter 3
Principles of Design Space Exploration

The major objective for system architects is to identify an optimal design point with respect to the main objectives of performance, flexibility, energy and power efficiency, and costs. This identification requires a structured design methodology defining the fundamentals of *design space exploration*.

- *Evaluation* of a single design point is mandatory to determine its quality with respect to the objectives.
- *Exploration* defines the search within the huge design space to find the optimal or near optimal implementation. As the design space is spanned by all the software and hardware design options, it prohibits a complete search. As a consequence, this demands a knowledge-guided search of the design space.

In general, exploration defines a multiobjective optimization problem [74]. The given constraints restrict the design space to feasible implementation options, whereas each possible design represents a System-on-Chip (SoC) solution. The key components of the decision vector are the *software* space, including algorithmic decisions and task-level partitioning aspects, the *hardware architecture* and the *temporal and spatial task mapping*, depicted qualitatively in Fig. 3.1. The complexity and size of the design space along with the time for evaluating a single design point, prevents an exhaustive search to find the optimal solution. Consequently, only part of the complete design space can be elaborated, which naturally results in suboptimal solutions.

However, the number of investigated design points closely relates to the time required for the individual evaluation process. This process consists of two major parts, given by the time spent in developing and describing the intended design point, as well as the time for finally evaluating the anticipated design. The first defines the *modeling efficiency*, whereas the latter depends upon the *analysis* and/or *simulation time* for the pure design evaluation.

These two issues, along with the rapidly increasing design complexity, have driven extensive research in the past, today, and most likely in the future. Therefore, this chapter provides detailed background information on the general principles and requirements for successful design space exploration. Together with the previously discussed challenges of MPSoC platforms, these form the foundation of the proposed design space exploration.

T. Kempf et al., *Multiprocessor Systems on Chip: Design Space Exploration*,
DOI 10.1007/978-1-4419-8153-0_3, © Springer Science+Business Media, LLC 2011

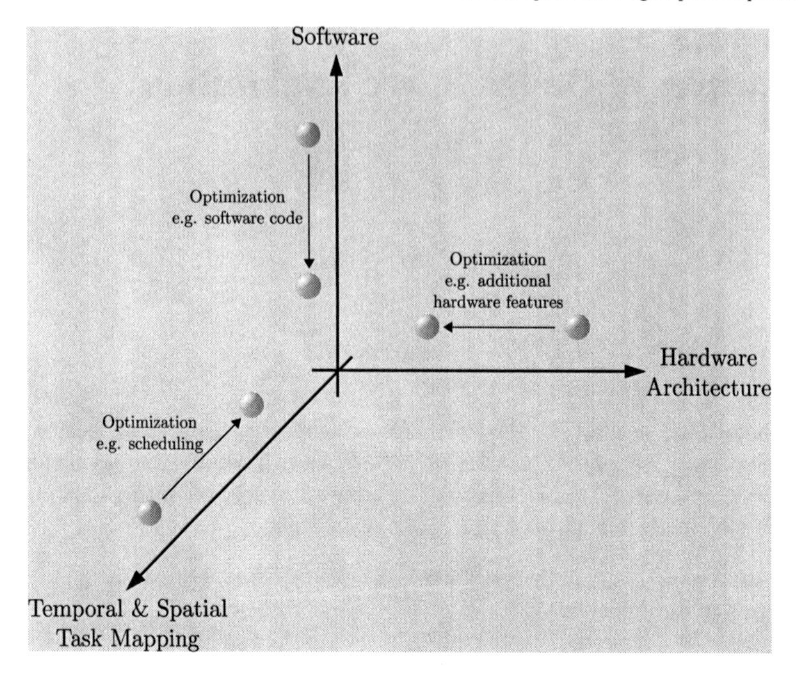

Fig. 3.1 Design space: software, hardware architecture, and task mapping

3.1 Evaluation of a Single Design Point

The evaluation of a single design point has always been a key element for design space exploration. The evaluation typically measures objectives like area, energy, and timing during exploration. Apparently, evaluating such properties at the level of a single IP component, e.g., a low-pass filter or an FFT component, is rather simple compared to the measurement of a complete MPSoC. To cope with the increasing complexity, the abstraction level of the design methodology has been constantly raised from full-custom design to transistor level and RTL. The basic principle of all approaches is the method of *clustering* and *abstracting* [42, p. 206]. *ESL design* is still in ongoing development (Fig. 3.2) and its final shape is yet unknown. Reference [42] probably gives the most comprehensive snapshot of the status of ESL design methodology in 2006. It describes ESL design as a patchwork of different methodologies and frameworks, all aiming at particular pieces of the design issues. In general, based on the classification of Bailey et al. [42], the ESL design flow defines eight key aspects: specification and modeling, prepartitioning analysis, partitioning, postpartitioning analysis and debugging, postpartitioning verification, hardware implementation, software implementation, and implementation verification.

The work described in this book cannot be simply reduced to only one of these aspects. Rather than that, it spans different aspects such as partitioning, postpartitioning analysis, and debugging, together with software and hardware implementation. The complete set of aspects defines a structured top-down design

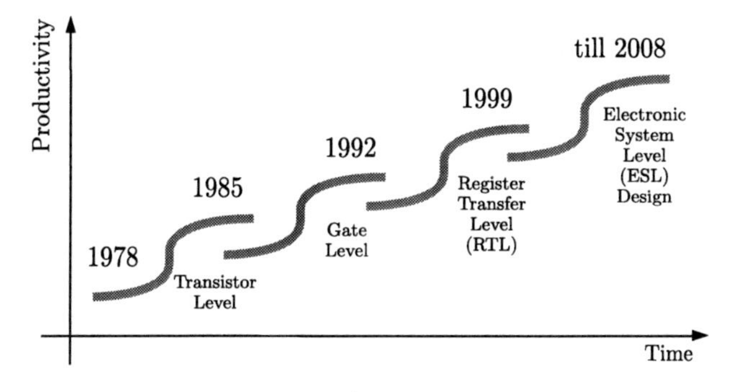

Fig. 3.2 S-Curves: abstraction levels of hardware design [75]

approach. Such approaches typically cover not only one level of abstraction but also many levels. However, for efficient and widely accepted design methods, an ordered refinement from high to low abstraction level is essential. Besides these top-down approaches, others focus on the principle of bottom-up, Y-chart [76], or PbD [33]. In theory these design methodologies are completely separate from each other, however in practice a mixture is used to opt for an optimal design and to keep design efficiency and IP component reuse high.

Gries defines design space exploration as a design funnel [77] of six abstraction levels. At early stages, high levels of abstraction allow for a fast exploration method to inspect a large region of the complete design space. Following the top-down principle, accuracy is increased step-by-step. This refinement goes hand in hand with increasing design effort, implementation, and evaluation time. The two highest abstraction levels, i.e., the system-level analytical model and abstract performance simulation, build the centerpiece of the proposed design methodology within this book. Additionally, the refinement to instruction-accurate and cycle-accurate simulation is the key for a smooth transition to the final implementation and is covered as well. Further refinement can rely on existing technologies. Therefore, the level of cycle-accurate instruction set simulation builds the bottom-line of the proposed approach.

The classic separation of evaluation methods leads to the two main concepts of *analytical* and *simulation*-based evaluation. For practical purposes a mixture of both is used for the evaluation of a single design point. Apart from these different approaches, the design entry typically differs significantly from one design to another. On the side of hardware architectures, the design entry can be the complete reuse of an existing platform, minor or major revisions of an existing one, or even a blank sheet (Fig. 3.3). Additionally, the design entry from the application point of view is characterized by various levels of knowledge. Figure 3.3 classifies the design entries with respect to the applications into three coarse-grained experience levels given by algorithmic and implementation knowledge, as well as the final implementation. In general, the design entries determine which approach, whether analytical or simulation-based, should be selected to achieve optimal exploration results. Commonly used simulation techniques are discussed later.

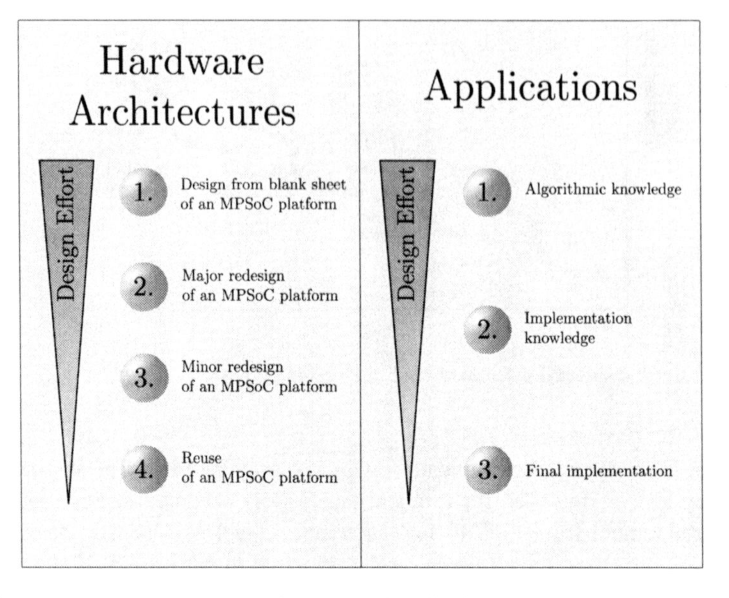

Fig. 3.3 Design entries: hardware architectures and applications

3.1.1 Simulation-Based Approaches

The IEEE defines a simulation model as:

> (1) A model that behaves or operates like a given system when provided a set of controlled inputs. (2) The process of developing or using a model as in (1) [78].

As this model and the measured execution characteristic heavily depend on the input (also called simulation stimulus [31]), their selection requires careful inspection to trigger the intended use-case. Apart from this issue, simulation models are well suited to identifying characteristics of a system in the presence of dynamic and hardly predictable effects, which impede or even prevent finding a simplified analytical model. Additionally, system architects can determine realistic performance parameters for average and worst-case scenarios if the right stimuli are selected.

The rapidly increasing complexity has been answered by constantly increasing the abstraction level and still maintaining and utilizing the lower abstraction levels. Figure 3.2 has highlighted the overall abstraction levels. Whereas in the 1990s hardware development mostly focused on RTL design, the level of abstraction has nowadays been raised to ESL design. In contrast to other abstraction levels, ESL design has not reached a final stage and its current status is much more a patchwork of methodologies than a single structured methodology [42]. Similar to the use of different abstraction levels like transistor, gate, or register-transfer level, the current ESL design includes different levels of abstractions. Unfortunately, there is no perfect abstraction level or simulation model for a complete system design. As expressed in the earlier-mentioned definition of simulation, each particular abstraction level serves a different purpose. For example, instruction-set simulation on a cycle-accurate level is mostly used for verification, whereas instruction-accurate simulations are mostly utilized for fast software development where only performance trends have to be measured.

Because of the different abstraction levels and utilized methodologies in system-level design, the design flow is a successive refinement of the simulation model from a high to low level of abstraction until the final implementation is defined.

According to [79] the design task can be separated into four orthogonal main aspects: *computation*, *data*, *time*, and *communication*. Software and hardware effects can be separately investigated for each main aspect. The computation characterizes the system's behavior. At the highest level it is defined by relations and constraints, whereas at the bottom level it finally resolves into transistors for hardware and into instructions for software. The time aspect deals with all timing issues from rough requirements down to the physical time. The data attribute varies from symbols that can hold arbitrarily complex data containers, such as data packets, to fine-grained continuous values that quantify physical units. Finally, the communication aspect considers the high-level structure and the interface constraints, while at the low-level layout issues are of interest.

Based on the objective of the simulation model, an arbitrary combination of different abstraction levels is typically used to focus on a particular issue of the design. Therefore, the refinement of one aspect is supported independently of the others. This leaves the overall platform at a given level of abstraction and refines a particular component toward the final implementation. Based on the different objectives of system models, a wide variety of *models of computations (MoC)* have been developed.

Model of Computation

Jantsch [79] defines a MoC as:

> A model is a simplification of another entity, which can be a physical thing or another model. The model contains exactly those characteristics and properties of the modeled entity that are relevant for a given task.

In general, modeling a system separates it into a set of components which interact with each other and the environment [80]. In addition to that, the MoC defines the component interaction and their underlying behavior.

When modeling heterogeneous MPSoCs a single model cannot support all kinds of issues, e.g., modeling the performance, functionality, and verification. A wide range of different MoCs can be found in literature, each addressing a specific design issue to keep the complexity for that particular model reasonable.

The Tagged Signal Model [81] provides a methodology to formally specify and compare timed and untimed models of computation. This modeling principle builds upon processes, signals, and tags to reflect the basic MoCs such as KPN [19], discrete-event systems [82], and reactive process networks [83]. Jantsch [79] further separates the synchronous models of computation from the timed ones as these operate on slots or cycle basis. In the following, a brief overview of these models of computation is given.

Untimed MoCs

The key identifier of untimed MoCs is the complete absence of timing information. Only causality is ensured by ordering the occurring events. Classic untimed MoCs have originated from dataflow process networks [19, 84] addressing the analysis of signal-processing algorithms. Further development of process networks has led to SDF graphs [20]. Thanks to their special properties, SDF graphs have been extensively utilized to solve important problems like computing a static schedule for single- [85] and multiprocessor systems [86] as well as finding optimal buffer sizes. Among them, further data-centric untimed MoCs are Boolean Data Flow graphs [87] and Process Coordination Calculus (PCC) [88].

Apart from these MoCs, others are more control oriented. Prominent examples that can be covered by the Tagged Signal Modeling are the ones based on the rendezvous of sequential processes such as Calculus for Communicating Systems (CCS) [89] and Communicating Sequential Processes (CSP) [90], which is the basis of the protocol specification language Specification and Description language (SDL) [91].

Synchronous MoCs

These can be separated into two major MoCs, the perfectly synchronous one and the clocked synchronous one [79]. The key principle of both is to divide the time into slots or cycles and everything within one slot or cycle takes place at the same time. Time progresses by ordered slots or cycles. In addition, perfect synchrony assumes that neither the communication nor the computation takes time, while in a clocked synchronous MoC computation is assumed to require a single cycle. Besides this partitioning, a further classification can again be performed on the basis of the data flow and control flow. Examples for data flow dominated ones are Signal [92] and Lustre [93], while prominent control-dominated approaches are Esterel [94] and Argos [95]. Since the foundation of all synchronous MoCs is well defined, formal verification methods have been successfully applied to them. This has led to the use of such approaches in the domain of control critical systems such as airplane controls.

Timed MoCs

In contrast to synchronous MoCs that restrict the time resolution to a single time slot, timed MoCs are much more general and allow event generation at any time and in arbitrary number. Additionally, each event is connected to a particular delay period, which keeps the notion of time. Timed MoCs are quite popular in the domain of hardware simulation. Classic simulators operate on the principle of Δ-cycles like VHDL [96] and SystemC [97].

Following one particular or combining multiple MoCs, many design environments emerged in the past. These have been consistently subsumed under the term HW/SW codesign and form the body of traditional design space exploration.

HW/SW Codesign and Traditional Design Space Exploration

The transition from board-level design to systems-on-chip has opened up completely new opportunities, but also challenges due to the joint utilization of software and hardware. More than a decade of HW/SW-codesign research activities have given birth to various design methodologies and tools. In this section only a brief overview is given, whereas in-depth discussions can be found in [98, 99].

In general, early approaches can be divided into four major directions: *HW/SW cosimulation*, *system synthesis*, *communication analysis and synthesis*, and *interface synthesis* [100].

To enable the joint development and debugging of both, software and hardware, research targeted *HW/SW cosimulation* frameworks. The general approach was to provide a simulation model for performance analysis and functional verification [101], that was available prior to the hardware prototype.

Among the first available frameworks, the *Ptolemy* [102] cosimulation framework combined execution capabilities for the most common MoCs. These include the MoCs of discrete-event, dynamic data-flow, and synchronous data-flow. During simulation, the Ptolemy framework coordinates them and provides efficient mechanisms for the interaction of the various domains.

Based on the Ptolemy simulator, the framework called *POLIS* [103] provided a unified methodology for jointly modeling hardware and software. The key idea was to model the system by codesign finite state machines allowing annotation of timing characteristics. As the modeling of state machines can be performed either in hardware or software languages, the concept was generic.

Taking the ideas of Ptolemy and POLIS, Cadence released the tool *Virtual Component Codesign* (VCC) [104] at the beginning of 2000. Despite wide interest in research and academia the tool was withdrawn in 2002 [42, p. 45f.]. During the same time period Synopsys released a similar modeling environment called CoCentric System Studio that is still active today and is one of the leading ESL design tools [105].

The initial concepts for the combined simulation of hardware description languages and instruction set simulators, like the still actively supported *Mentor Seamless* [106], rapidly found interest in industry. Since the combined execution limited the simulation speed of such approaches, the *GRACE++* methodology [107, 108], among others, targeted abstraction levels higher than RTL to achieve increased simulation speed. Yet, modeling was still cycle and pin accurate.

In contrast to HW/SW cosimulation, methodologies based on synthesis approaches aimed at an automatic generation of the system starting from a formal description. Full *system synthesis* was the most ambitious goal targeted by some approaches. The most notable ones are *Vulcan* [109], *COSYMA* [110], *LYCOS* [111], and *Cosmos* [112]. None of them has found a large community and relevance with respect to the intended goal. However, the semiautomatic approaches of communication and interface synthesis emerged from these fundamental techniques.

With the rising number of applied IP components within a single MPSoC design, the basic objective of *communication synthesis* is to generate the communication architecture and infrastructure automatically for a given application. Early attempts used static analysis and synthesis techniques [111–113] leading either to too optimistic results or allowing only investigation of worst-case scenarios. More advanced approaches like the one proposed by Lahiri et al. [114] combine simulation and analysis results to increase the modeling accuracy.

Most of the approaches are limited to simple communication architectures like point-to-point or bus-based communication. Addressing more complex architectures like Networks-on-Chips requires manual interaction. Nevertheless, these techniques support and improve the design process like the *NetChip* [115] project based on the underlying *X-Pipes* [116] and *SUNMAP* [117] technologies.

The forth and final major direction of HW/SW codesign is the *interface synthesis*. The general concept only aims at supporting designers in their decisions and not to make any automatic synthesis of parts or the complete system. Hence, it relieves designers from repeating tasks, finally reducing the design effort and occurring errors. The most notable example of interface synthesis was initially developed at IMEC and found its way into CoWare's products [118].

The basic concepts and ideas of later *System Level Design* have been envisioned on the foundation of HW/SW codesign techniques. While in early HW/SW codesign phases vast effort was put into finding *the* model of computation, in system-level design focus has shifted to generic modeling concepts and languages. Early attempts culminated in the ESL design environments introduced in the following.

Based on the approach of *POLIS*, the design space exploration framework called *Metropolis* [119] was developed. Metropolis removed the modeling demand of codesign finite state machines which has been considered to be the major obstacle for utilizing the POLIS framework in practice. The replacement for these state machines was the Metropolis Meta Model [120] which was still rather formal, achieving the key benefits, but keeping the modeling less strict.

The system-level performance analysis and design space exploration (*SPADE*) [121] project also has its roots in early HW/SW codesign. SPADE was among the first frameworks that followed the y-chart principle and applied various abstraction levels [122]. The fundamental technique was based on KPN [19] for which the programming interface YAPI [123] was envisioned.

The successor of SPADE called *ARTEMIS* [124] extended the framework into two separate directions. The first, covered by the *Sesame* project [125], improved the simulation accuracy by including fine-grained architecture models [126].

In addition, automatic design space exploration techniques were incorporated [127]. Following an alternative approach, the *Archer* project [128] replaced application modeling based on KPNs by symbolic programs closely related to control data flow graphs. Recently, the activities within Sesame have been incorporated into the *Daedalus* project [129] trying to bridge the gap from system-level design based on the Sesame methodology to final RTL implementation.

Another approach based on KPNs is the Philips' *Eclipse* [130] platform and framework. Limited to computationally demanding multimedia processing, the anticipated applications are efficiently composed out of KPNs. An optimized application-to-architecture mapping extensively uses specialized communication architectures, e.g., hardware FIFOs, to implement the occurring FIFOs within the KPN.

The *Modeling Environment for Software and Hardware* (MESH) [131] project raises the abstraction level from cycle-accurate modeling. It provides different techniques to capture hardware effects, like complex processing elements and communication architectures [132], by considering schedulers and scheduling operations as the central element [133].

The *ARTS* [134] simulation framework proposes a representation to jointly evaluate the complete design space spread by application, architecture with the dominant components of processing elements and communication architectures. Further extensions and conceptional proofs discussed the underlying architectural models [135, 136] and software impact [137].

Another approach has been initially developed for ST Microelectronics *StepNP* [138] platform. The framework represents a development environment including the design aspects of architecture, application, and tools. Targeting architectural design, a construction library contains the most dominant parts for assembling complete platforms. For the software domain, the approach includes the *Click* [139] framework which allows simple and efficient software development according to a component-based approach [26]. Furthermore, the *NP-Click* programming model has been developed based on Click for the specific requirements of network processors [140].

The *ROSES* [141] framework addresses IP CbD [32] of hardware platforms. The major benefit is the reduced design effort and the quick design composition and investigation. Further proposed extensions include mechanisms to automatically generate interfaces for software and hardware.

At the beginning of HW/SW codesign and system-level design frameworks, research focused on the underlying MoC and a formal description. As no final MoC has been found serving all kinds of different aspects, the only well accepted MoC for hardware simulation is the discrete-event model utilized in VHDL, Verilog, and SystemC simulators. Although VHDL and Verilog are mostly used for RTL design and below, SystemC has recently gained wide acceptance in the area of system-level design. Other ESL languages like SpecC [142] seem to be disappearing at this point in time.

ESL Design Languages

The results of early HW/SW codesign and the identified need for a suitable modeling language, attracted research groups worldwide. The first goal of all the proposed languages was to replace VHDL and Verilog as the dominant hardware description languages, adding the capabilities of system-level modeling. However, until today none of the approaches has achieved this ambitious goal and SystemC, as the dominant language, has abandoned this first objective.

The Open SystemC Initiative (OSCI) [143] has been founded and started creating the SystemC language in 1999, at the same time the SpecC Technology Open Consortium (STOC) [144] developed the SpecC language. Originally designed at U.C. Irvine, the SpecC language proposed constructs like abstraction, orthogonalization of concerns, clear defined interfacing, and increased design reuse. There are several reasons why SpecC has completely lost ground with respect to SystemC. One reason is that SystemC is built as a C++ library, which inherently supports all the well-known software constructs like Object-Oriented Programming and has a large developer community. Another reason was the absence of all big commercial EDA vendors in the SpecC consortium, whereas OSCI included the big players [42].

Unlike the top-down approaches of SystemC and SpecC, the proposed modeling language of SystemVerilog [145] followed a bottom-up strategy. Taking the hardware description language Verilog [58] as entry point, the language has been extended by concepts from OpenVera [146] and Superlog [146]. Despite some controversial opinions about what became the true system-level language, in 2004 it was agreed that both are complementary. After that, SystemC became the dominating system-level language, while SystemVerilog jointly comprised a hardware description and verification language. A more detailed discussion referred to as the *system-level language war* can be found in [42, p. 54f].

SystemC is built upon C++ as a class library that enables hardware modeling by introducing concurrency and time. The underlying generic model of computation is based on the interface method call (IMC) principle [147] that is reflected by the discrete-event simulation kernel of SystemC. In addition, the principle of IMC supports the separation of communication and computation in terms of an orthogonalization of concerns [148]. For a detailed discussion of these fundamental properties of the SystemC language the reader is referred to the book *System Design with SystemC* by Grötker et al. [97].

Today SystemC has gone through several revisions and the basic SystemC kernel has advanced to version 2.2. Apart from the pure kernel, additional libraries and methodologies have been provided by different OSCI working groups. The most prominent enhancements are the Transaction Level Modeling (TLM) Library in release 2.0 and the SystemC Verification Library (SCV). Based on the solid foundation of the TLM library, efficient modeling and development of complex systems are possible. Hence, the subsequent section briefly discusses the benefits of TLM.

Transaction Level Modeling

The fine-grained pin and cycle-accurate modeling at RTL was incorporated in early system-level approaches. Soon it became one of the major drawbacks for efficient design and exploration of systems. Besides the low simulation speeds, the effort in coupling components was considerable and prohibited easy and fast exploration of various design decisions. To overcome these issues TLM was envisioned on the principle of IMCs [147].

The introduction of the TLM 1.0 standard had a significant impact on the community at that time. It supported a structured methodology for different use cases of system modeling (untimed, approximately, and cycle-accurate timed) and already contained several primitives for sending and receiving data (Fig. 3.4). However, these data tokens were kept generic and were not defined by the TLM standard. This resulted in a wide variety of interface implementations. Early attempts like VSIA [150] to unify them failed. Therefore, the defined interfaces were and still

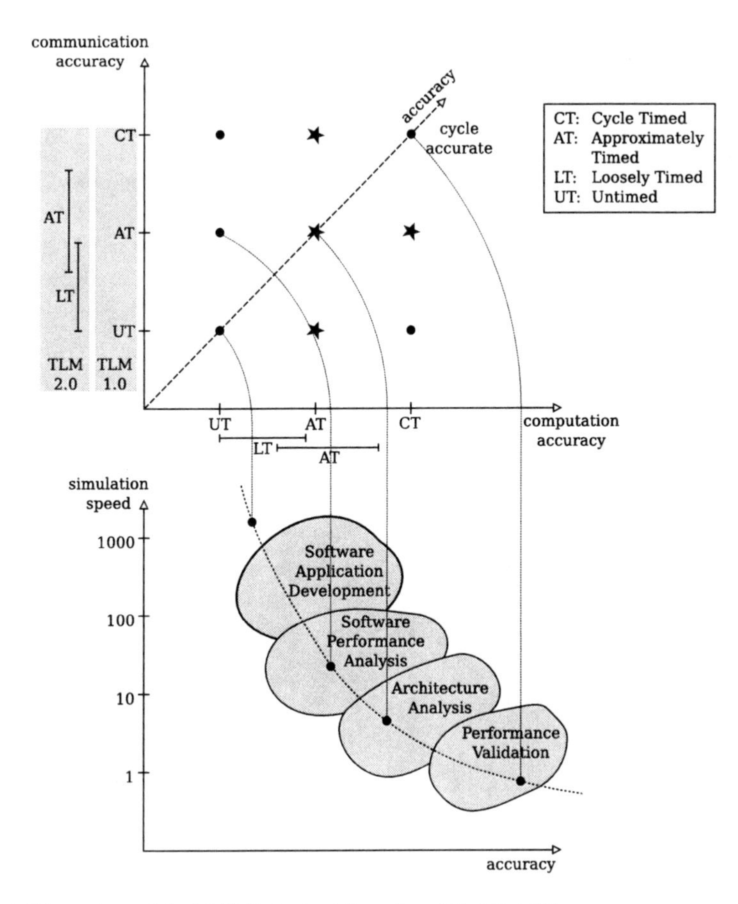

Fig. 3.4 Dimensions of the TLM-2 standard (based on [39, 100, 149])

are dominated by the underlying communication architecture interfaces, e.g., the AMBA buses including AHB, APB, and AXI [34] as well as IBM's PLB and OPB [35].

In addition, the nonprofit Open Core Protocol International Partnership (OCP-IP) [66] was founded in 2002 to define a standardized communication interface with the envisioned goal of getting compliant IP components. According to the TLM standard, the OCP-IP consortium defined three transaction levels (TL) above RTL [151]. Although TL-1 (transfer layer) of the OCP-IP standard defines a cycle- and word-accurate representation, TL-2 (transaction layer) abstracts the complete transaction to, e.g., a single abstract blocking function call, and finally TL-3 (message layer) aims at generic architecture exploration where timing can be annotated at data-packet level. A detailed discussion of all layers can be found in [152].

With APIs being proprietary in the past, the recently introduced TLM 2.0 (TLM-2) standard [153] addresses this issue, among other technology improvements. The key objective is to increase the interoperability between the different available tools and models.

Figure 3.4 depicts the dimensions of the TLM 2.0 standard. The lower part of the figure reflects the trade-off between accuracy and simulation speed. The working group has defined four major use cases: *software application development*, *software-performance analysis*, *architecture analysis*, and *performance validation*. Each use case operates in a particular area as qualitatively depicted in the figure. The upper part highlights the defined levels of computation and communication accuracy. The combination defines the overall accuracy measure (lower part of Fig. 3.4).

Instead of defining a fixed modeling set for each and every use case, the standard provides the possibility to arbitrarily select different modeling styles and mechanisms (lower part of Fig. 3.5) for an intended use case (upper part of Fig. 3.5). A detailed discussion is found within the TLM-2 standard [153].

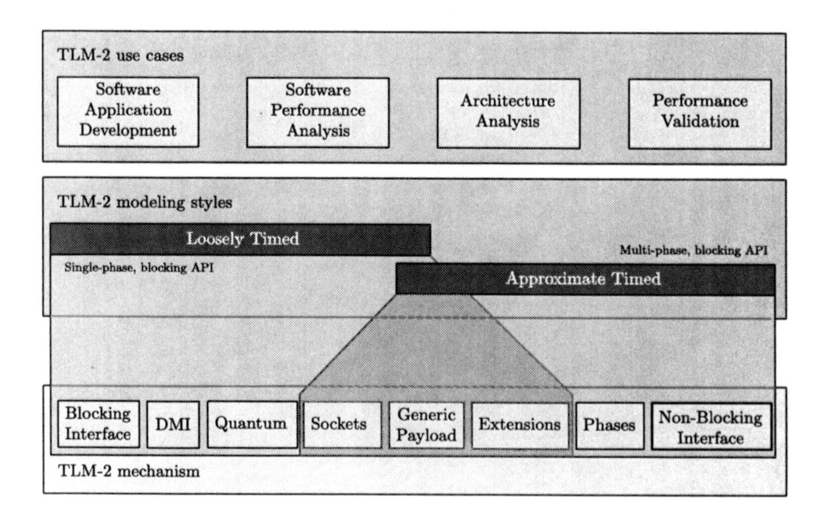

Fig. 3.5 Use cases, modeling styles and mechanisms [153]

Unfortunately, no unique simulation model exists that serves every purpose. Instead, as mentioned before a set of major use cases has been defined, that cover the most critical design issues. The important use cases and their potential [42, page 42f.] can be summarized as follows:

- *Software development and debugging.* Development and debugging of software can easily start in advance before the hardware prototype is available, which shortens the design cycle significantly. In addition, deterministic simulation behavior allows the reproduction of the encountered software errors and simplifies debugging, due to the complete visibility of internal states and thanks to the full simulation control. Especially, the latter is typically not available in hardware prototypes [154].
- *Software performance analysis.* Contrary to RTL-based simulation models, virtual platforms allow the investigation of various complex software and hardware scenarios, like booting an operating system at acceptable speed. The key enabler is the high simulation speed, which is typically three orders of magnitude faster than RTL-based simulation [39].
- *Architecture analysis and optimizations.* Avoiding long and cost-intensive hardware prototyping, virtual platforms allow efficient hardware architecture optimizations by replacing most of the hardware prototyping activities. This reduces design costs and speeds up the design process significantly. The architectural objectives range from coarse-grained system-level aspects to the fine-grained selection of custom instructions for a single processor core.
- *Performance validation.* Validation of the system performance requires cycle-accurate virtual platforms. Because of detailed modeling, such validation must accept reduced simulation speed. Moreover, the ability to incorporate RTL cosimulation allows developing efficient testbeds for functional verification, avoiding error prone reimplementation.

The TLM-2 standard serves as a background technology for the efficient design of simulation models, shielding developers from low-level hardware interface modeling. The major advantage is the smooth integration of simulation models of various IP components.

Based on the solid foundation of SystemC and TLM, a new class of simulation models called virtual platforms has emerged. These are an exact image of the intended hardware platform within a simulation environment. Accepted as a major technology breakthrough, the following discussion highlights these together with the available design environments.

Virtual Platforms and Design Environments

Virtual Platforms (VP), formerly called virtual system prototypes (VSP) [155], define behavioral models of a particular platform at various abstraction levels. Typically, they are utilized as an executable specification. The value can be enormous when applied carefully. The most considerable capabilities are:

- Software development and debugging prior to the hardware prototype are possible. Nonintrusive software debugging is even further enhanced by better state visibility, reproducible scenarios, and full simulation control.
- Analysis, optimization, and verification of a complete platform including hardware and software.
- Easy exploration of design alternatives, which avoid cost-intensive hardware re-designs.

Summarizing the key advantages, the overall design time is reduced significantly and the technology provides high potential to design more cost-, performance-, and energy-efficient hardware platforms.

In addition, modern platform-design philosophies, like CbD [32] and PbD [33], have a strong influence on the paradigms of virtual platforms and the underlying TLM.

- Following CbD, a platform consists of various IP components connected to each other. These components communicate over well-defined interfaces (also referred to as ports) and the interior of any component is a black box for the others.
- PbD assumes one of such composed platforms as a base architecture. This base architecture serves as a foundation for the development of various market- and application-specific architectures. In general, this requires incorporating additional IP components to address specific needs.

These two design principles inherently ensure structured platforms given by the assembly of IP components. Accordingly, the paradigm of TLM follows this philosophy. The clear separation of components (SystemC models) represents the structure of IP components within the ESL domain based on the paradigm of orthogonalization of concerns [156] and the TLM-2 standard [153].

The efficient modeling of the individual IP components is highly important. SystemC and the TLM-2 standard provide the basic set of modeling primitives and techniques to model any IP component. However, the OSCI standard itself defines only the modeling method and provides rather simple examples to demonstrate the envisioned capabilities. Nowadays, large component libraries [157,158] exist, which are compatible with the TLM-2 standard. In addition, various SystemC modeling tutorials and guidelines are available [97, 159].

SystemC provides efficient techniques for modeling communication architectures, peripherals, memories, and hardware accelerators. However, modeling the most important IP-component type, i.e., *processor cores*, has always been the most complex assignment which requires special treatment. Processor cores are always defined by hardware and software together, whereas fixed hardware components execute the different functionalities in a deterministic order. Accordingly, modeling and simulating processor cores have always been a focal point and many simulation techniques emerged in the past. The common principle is based on instruction set simulation [160], but the underlying techniques differ significantly and show many facets and variations.

Instruction Set Simulators

Various techniques have been proposed to support efficient and fast modeling and simulation. Wieferink et al. [161] give a thorough discussion on the various modeling schemes and propose a successive refinement flow. Within the main contribution, instruction set simulators are classified into different classes located above traditional RTL modeling. Separated into the two main classes of *instruction-accurate* and *cycle-accurate* simulation models, further subclasses have been defined by the utilization of interfaces like bus cycle-accurate or more abstract functional interfaces.

Another classification, in [162], separates instruction set simulators according to their underlying simulation techniques, including interpretive simulation as well as statically and dynamically compiled approaches. One of the key challenges was and still is to achieve high simulation speeds. Hence, after basic interpretive simulation techniques, statically compiled approaches appeared next. The general technique is to translate the binary executable of the simulated processor core into the binary format of the host processor [163, 164].

Statically compiled simulations can achieve superior performance due to joint compilation of software and hardware, but there are also certain drawbacks. A main limitation is that self-referential or self-modifying software code cannot be handled and each software modification requires time intensive re-compilation of the simulator and most likely the complete platform model. To overcome these issues, dynamic compilation techniques nowadays dominate the field of instruction set simulators. They are able to directly simulate modified software without any time intensive re-compilation steps while achieving reasonable simulation speeds. To speed-up dynamically compiled simulation, several techniques, like Just-In-Time cached compilation [165–168] or Just-In-Time binary translation [169], have been incorporated. Related approaches can be found in the domain of software instrumentation techniques [170], in particular in dynamic binary instrumentation (DBI) frameworks.

Based on the principle of orthogonalization of concerns, each component treats every other component as a black box and data exchange only occurs at the well-defined interfaces and ports. Hence, every instruction set simulator can simply be encapsulated within a SystemC component. By assembling all platform components, a simulation model is obtained commonly referred to as Virtual Platform (VP).

Virtual Platforms

Several vendors approach the upcoming market segment with a large variety of environments and tools. In addition, a few open-source solutions are available. After the introduction of commercial tools, the open-source approaches are highlighted.

Platform Architect [171] of CoWare is one of the most well known design environments for SystemC development and simulation. In contrast to a single design tool, Platform Architect defines a complete tool suite addressing the requirements of the multiple design challenges. The SCIDE tool, based on the Eclipse IDE [172], allows for SystemC debugging, while the Platform Creator Tool (PCT) [173] provides a graphical design entry for efficient platform design. Platform Architect comes with a rich set of libraries containing the most common IP components, e.g., ARM processor cores [174] and AMBA buses [34]. The Architects View Framework (AVF) library [100] enables efficient exploration of complex communication architectures. The underlying technique of the AVF is based on the Virtual Architecture Mapping (VAM) [175].

Another well-known design environment is Synopsys *Innovator* (formerly Virtio) [176]. In the past it was based on a proprietary design technology based on C/C++, but nowadays it is fully compliant with the SystemC standard. Similar to CoWare design tools, it provides an environment for efficient development, simulation, and analysis of virtual platforms. In addition to the Innovator product, Synopsys offers the DesignWare IP component library [157] and System Studio [105] as a graphical design tool for platform design.

Carbon Design Systems has recently acquired the virtual platform development suite *SoC Designer* [177] from ARM. In contrast to CoWare Platform Architect, SoC Designer focuses on cycle accurate virtual platforms. This smoothly links to other Carbon technologies, e.g., Model Studio, that allow automatic extraction of SystemC models from low RTL code.

Besides these, other commercial tools exist, like *Simics* from Virtutech [178], Vast *CoMET* [179], and *Triton Tuner* [180].

Recently, Imperas Inc. has donated its environment as *Open Virtual Platform* (OVP) [181] to the open-source community. Additionally, a GreenSocs [182] project called *VPP* targets another open-source approach.

Formed by ST Microelectronic, the *On-Chip Communication Network* (OCCN) [183] project aims to develop an open-source interface based on the SystemC kernel, targeting complex communication architectures [184]. The main contribution is a layered protocol stack including a well-defined methodology for protocol refinement.

Apart from the simulation-based approaches, analytical ones target the exploration and evaluation in a much more formal way.

3.1.2 Analytical Approaches

Analytical approaches are applicable directly at the start of the design. Therefore, they are well suited to speeding up the exploration and to identifying *corner cases* for later system simulation. The general assumption for most of the formal methods is that nondeterministic and dynamic behavior is prohibited. In addition, these methods typically address worst-case execution time (WCET) evaluation, especially at early design stages.

These analytical approaches operate on a coarse-grained evaluation based on:

- The hardware architecture.
- The application partitioning.
- The application-to-architecture mapping.

The problem of mapping an application onto a multiprocessor platform is a well-known issue in computer science. Depending on the point of view, this problem can be found in literature as *scheduling*, *resource allocation*, or *task mapping*. In general, it is known to be part of the class of NP-hard problems [185] as well as its simplified subproblems [186]. The common classification separates the overall problem into two categories, static and dynamic.

Both techniques define the order of task execution on a shared resource and have been extensively studied in the past. When utilizing a *static schedule*, the task execution order is fixed and computed before run-time, whereas a *dynamic scheduler* determines the order at run-time. Hence, the scheduling overhead is typically larger in the presence of a dynamic scheduler, but a priori knowledge of the execution behavior is required to compute an efficient static schedule. Reference [187] discusses the known scheduling algorithms separately for the classes of graph theory, mathematical programming, queuing theory, and finally solution space enumeration and search.

Originating from queuing theory of computer networks [188] and linear-system theory [189], the technique of *Network Calculus* [190] has been adopted for the purpose of performance evaluation of embedded systems. The main idea proposed by Thiele et al. [191] abstracts the application to arrival curves, whereas the platform resources are captured by service curves. When computing the characteristic of the application mapped onto the given platform, both curves are joined and the results illustrate the resource utilization and the execution performance. The principle of network calculus has been successfully applied to the design of network processors [192–194] and more general embedded systems [195].

Following a similar approach, Richter et al. operate on *workload models* [196] to allow the symbolic calculation of execution characteristics. Furthermore, this approach has led to the SymTA/S approach [197] that has been integrated into a commercial product from Symtavision. For a detailed discussion of this and the network calculus approach [198–200] should be consulted.

Apart from the rather roughly computed bounds, the development of analytical models with a certain degree of precision can be extremely difficult and can take significant development effort [201]. In addition, these models do not allow any functional development and typically do not provide a design flow leading to the final implementation.

Other approaches [202] analyze a system based on *conditional process graphs* [203, 204]. Here, the application is described by a set of tasks mapped to a specific architecture allowing the computation of the performance characteristic. The goal of this work is to analyze and optimize the scheduling effects within heterogeneous systems including application and hardware-architecture effects.

Models based on SDF graphs [20] have been applied to single- and multiprocessors scenarios [86]. Both use cases have been studied extensively in research and have shown promising results in the evaluation of a single design point, e.g., in [205].

3.1.3 Joint Analytical and Simulation-Based Approaches

Combining both previously discussed techniques has mostly been used to inspect memory and cache behavior in the past. In particular, *trace-based performance analysis* has been extensively utilized to save simulation time by doing costly performance evaluation, e.g., of caches, once and then reusing the obtained traces in future. Generating and storing the trace allows inspection of various memory and cache architectures without any further trace generation. Memory exploration based on traces has been demonstrated by several case studies for memories [206–208] and cache performance [209].

Lahiri et al. [210] have extended the scope of *trace-driven analysis* to the design space exploration of communication architectures. Here, traces are measured once and taken for the computation of performance and resource utilization. Case studies and other approaches [211] have demonstrated the modeling accuracy even for complex buses and Network-on-Chips (NoCs) [212]. Further research has even broadened the investigated communication impact to evaluate the cost of control operations [213]. Another approach by Bobrek et al. [214] investigates resource contentions.

Apart from trace-driven approaches, other techniques require an *initial calibration* [215]. First, the measurement of a large set of benchmarks is performed to try to exhaustively cover the design space. In a second step, the retrieved information is taken to analytically evaluate performance and even power figures [216].

Another approach couples the network calculus and its event streams [199] with a SystemC simulation environment [217] by providing converters. The key element is the converter that outputs SystemC events from an input given as an arrival curve. For the sake of completeness, an inverse converter exists that generates arrival curves from SystemC events.

Summarizing the sketched approaches, most of them operate in two phases. In the first design phase, the execution characteristic of one or multiple benchmarks is measured. In a second step, the obtained information is used to analytically evaluate the optimized system-level design.

3.1.4 Summary of Approaches

Each of the three introduced methods to evaluate a single design point has its own advantages and disadvantages. Analytical approaches can speed-up the exploration of different design points and can help to find corner cases. However, they have the following common drawbacks.

- The limitation to worst-case or deterministic behavior might not always be sufficient for a precise investigation of complex systems. In addition, analytical models can hardly keep track of interactive and dynamic user behavior, which requires simulation-based approaches.
- Analytical models commonly lack real implementation, therefore, in addition to the time-consuming process of analytical modeling, the final software and hardware have to be developed separately in a different environment.
- The development of analytical models with a sufficient degree of precision can be extremely difficult and can take significant development effort [201].
- Due to the high abstraction level, a certain degree of inaccuracy and error must be accepted when compared against the real hardware implementation. Typically, this is not acceptable for final verification, demanding virtual platforms and even lower-level simulation models.

To overcome these limitations and issues, other approaches follow a pragmatic path by using simulation techniques. These simulation-based approaches support mixed abstraction levels and utilize different types of architecture and application modeling. Today most prominent ESL design frameworks are based on SystemC [143]. Such frameworks typically operate on cycle- and/or instruction-accurate level by the technique of instruction set simulation. Recently, virtual platforms have originated from such frameworks, focusing on simulation speed-up to allow software development and analysis before the final hardware is available. Virtual platforms have made a significant impact on the interoperability of design environments and modeling styles. They are now widely accepted as the current standard technology for software development and analysis of embedded systems. Prominent environments are CoWare Platform Architect [171], Synopsys System Studio [105], Vast [179], and Virtutech [178] among others. Unfortunately, for design space exploration at a particularly *early* design stage these frameworks suffer by nature from the following issues.

- *Hardware and Compiler.* To utilize the technique of instruction set simulation, a fixed processor core or at least a preliminary instruction set architecture has to be defined. Otherwise the development of an instruction set simulator or compiler tool-chain, including an assembler, cannot be performed at all. When addressing *early* design stages, typically none of them is present, which prevents performance evaluation until the processor core as well as the software tool chain is ready, which is far too late in the design process.
- *Software.* Optimized software development can only start jointly or after the hardware and compiler are fixed. Apart from this, the results obtained by instruction set simulation only measure the current design stage of software and hardware. Accordingly, these can only illustrate a current snapshot of the design. When entering a design process typically neither the software nor the hardware is mature and, if it exists, it can only be considered a basic reference

implementation. Measuring such intermediate stages does not reflect the final implementation, and design decisions based on such ad hoc measurements induce high risks of taking false decisions. Because the selection of a processor core at early design stages might imply a specific software development technique, a wrong decision can result in significant increased design effort. For example, it can induce the necessity of software rewriting or applying hardware-specific software optimizations twice. With short time-to-market, such redesigns typically have disastrous effects on the final product and its business success.

In the past combining both analytical and simulation-based approaches has mostly been performed to inspect memory and cache behavior as well as communication aspects. Other approaches utilize an analytical approach with an initial calibration based on simulation results or try to trigger simulation events from an analytical model. However, these approaches are separated from the design flow and cannot be integrated into a smooth design flow with successive refinement to the final implementation. In addition, these design environments are mostly completely decoupled from common simulation-based approaches like virtual platforms.

After the introduction of the evaluation techniques for a single design point, the discussion now turns to the different exploration techniques to navigate through the large and complex design space.

3.2 Exploring the Design Space

Design space exploration in the domain of logic synthesis originally refers to exploring different design options with respect to given constraints. This design problem can be comprehensively described as a multiobjective optimization problem [74], also known as Pareto optimization problem [218]. Sufficiently precise description of a classical integrated-circuit (IC) design defines such a problem. In general, area, timing, and energy are the main targeted system properties, however costs in terms of development effort can also be taken into account. Searching for an optimal solution does not lead to a single design point, rather the problem defines a set of optimal points in the design space also called Pareto-optimal or Pareto-efficient solutions.

To highlight this optimization problem, a small example from the domain of logic synthesis is consulted. Given the two system properties area and timing, the AT-product serves as optimization function ($f = A \times T$). In contrast to a single solution, an optimal AT-product defines a set of solutions as depicted in Fig. 3.6. The hyperbola $f_{\text{opt}} = A \times T$ or the line in double logarithmic scale (Fig. 3.6) defines a set of Pareto-optimal solutions, e.g., the illustrated designs 1–5.

Besides this simplified AT-product, the area–timing–energy (ATE) product is often considered in ASIC designs. The measurement of one or all the properties is mostly carried out by utilizing well established Register-Transfer-Level synthesis design flows like those of Synopsys [219], Cadence [220], or Magma [221].

For complete MPSoC platforms, even if the overall properties are still defined in the same way, measuring them gets far more complex than for a single hardwired

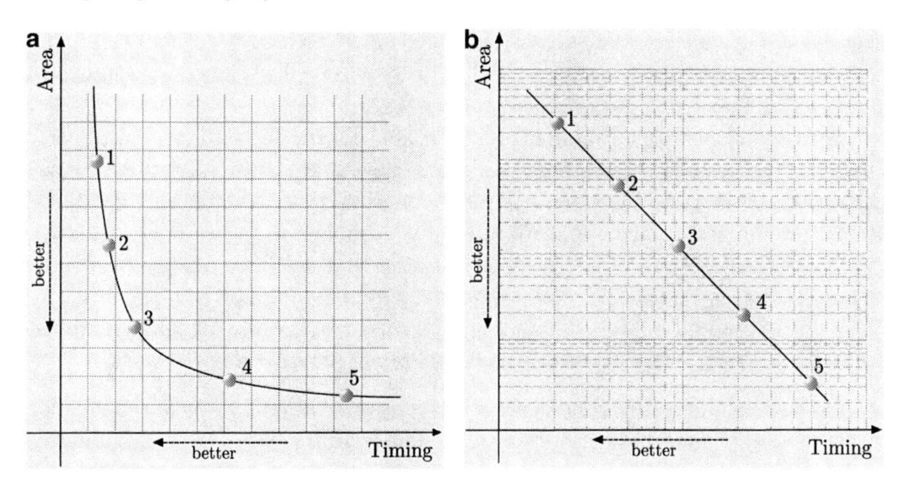

Fig. 3.6 Example for Pareto optimization: solutions for a minimal area-timing-product. (**a**) Linear scale. (**b**) Double-logarithmic scale

component. For example, the clock period and the execution time define the timing constraint for a single IP-component. When considering a complete MPSoC, the timing constraint reflects the execution of various interacting applications under arbitrary conditions. This makes it hard to determine the characteristic of even a single design point. In addition, system architects cannot directly manipulate these properties as they result from design decisions taken during the implementation of the application, which includes software and hardware.

The complexity of measuring a single design point (Sect. 3.1) and the sheer size of the design space prohibits an exhaustive search. To demonstrate this, a simple example is sketched. Assume an application consists of T tasks which can execute on n_{PE} processing elements. There are $(n_{\mathrm{PE}})^T$ possible design points. As typically the partitioning of the application and the selection of the hardware components are key points of the design space exploration, neither the number of tasks T nor the number of processing elements n_{PE} is fixed. This implies even more possible design candidates which cannot be evaluated exhaustively.

From theory, a large variety of different exploration strategies is known, e.g., random sampling and path-oriented search among many other techniques. A thorough introduction of these techniques can be found in [30, p. 148f]. However, as previously mentioned, the complexity of evaluating each design point along with the large number of possible design decisions makes it difficult to use such formal strategies.

This has led system architects to adopt more pragmatic design techniques like PbD [33] and IP component-based design (CbD) [32], which were introduced in Sect. 2.2. These paradigms guarantee high reuse and quick assembly of new platforms. Consequently, the design of new MPSoC platforms usually starts from a previous design that is extended and assembled out of existing IP components rather than designing the platform from scratch. This process reflects more a constant *evolution* rather than a complete re-design (Sect. 2.2).

As a consequence, the overall exploration can be separated into two linked parts. First, the fine-grained exploration of the individual components. Second, the coarse-grained exploration on the system-level where arbitrary components are combined to execute the anticipated applications jointly. Although in theory these two steps are independent, *only* the interwoven behavior can give evidence whether the overall system is working properly or not. Therefore, a joint exploration is mandatory.

As decisions on both levels have a significant impact on the overall system performance, thorough investigations are mandatory to prevent false decisions.

- Considering the *application* aspect, the key element is to partition the application into multiple tasks and to select the algorithms within them. The software implementation of a selected algorithm often depends on the anticipated processing element executing this task. For example, a Fast Fourier Transformation (FFT) can be implemented as radix-2, radix-4, or differently. While system architects partition the application and select the underlying algorithms on the system-level, the underlying implementation is chosen at the component-level.
- With respect to the *hardware*, the exploration selects the necessary components in terms of processing elements, communication architectures, and memories. On the system-level coarse-grained decisions are taken whether to incorporate one or the other processing element. Instead, on the component-level, fine-grained issues are determined. For example, when adding an ASIP the instruction set needs to be defined within this level.
- The application-to-architecture *mapping* has significant impact on the overall performance. In general, it temporally and spatially maps application tasks onto the underlying processing elements. Temporal mapping defines the execution order over time, whereas spatial mapping implies the distribution of tasks to processing elements.

These three aspects build the centerpiece during design space exploration for complex heterogeneous MPSoC platforms and apply to all abstraction levels.

With focus on the ESL design, in particular on virtual platforms including instruction set simulation, design space exploration is commonly carried out based on the *expert knowledge* of system architects and tools envisioned on the *workbench* paradigm. In this context, a workbench defines a tool supporting developers in taking decisions instead of trying to identify the best solution automatically.

This results in a design space exploration technique following the basic philosophy of implementing, measuring, analyzing, and modifying a possible platform candidate. Hence, at first system architects assemble the envisioned MPSoC hardware platform and software, thus creating a complete simulation model. This model is executed under different stimuli [31] and the performance is measured. The measurement results are then analyzed to check whether the constraints have been met or not. Depending on the result, either other design options are explored or the refinement toward the final implementation is targeted.

In general, all the earlier proposed ESL design techniques operate according to the *workbench* principle because of the immense complexity of heterogeneous MP-SoC platforms. A commonality among them is the use of instruction set simulation to mimic the behavior of processor cores.

Besides these workbench approaches, automatic-exploration tools have not yet found practical relevance and are only consulted in specific parts of the design, e.g., within the memory subsystem [30].

3.2.1 Summary of Exploration Approaches

The principle of design space exploration for MPSoC platforms considers the three main aspects of application, hardware architecture, and application-to-architecture mapping. Following the paradigms of component- and PbD, exploration is performed on two linked, but separate levels. On the *system-level* coarse-grained decisions about the overall structure of the platform are taken, whereas fine-grained design options are selected on the *component-level*.

In ESL design, the exploration follows a simulation-based approach with a *workbench* character. Here, system architects are supported by the methodology and tools to select the right design decisions. However, these design decisions need to be taken by system architects based on their expert knowledge because currently *no* overall automatic exploration technique exists.

3.3 Requirements for Early Design Space Exploration

So far the fundamental problem of heterogeneous MPSoC design and the related well-established methodologies have been introduced. The basic design problem is to determine a suitable application partitioning, a hardware architecture, and finally the application-to-architecture mapping. In general this comprises two main stages, i.e., the step of evaluating a single design point and the exploration step to find the best implementation candidate. In the context of evaluation, classical design methodologies utilize simulation-based approaches that are centric to instruction set simulation techniques. Additionally, analytical as well as joint analytical and simulation-based approaches can be utilized to measure the performance characteristics of a particular design point.

Evaluation techniques are essential to efficiently explore the design space and to finally identify a suitable software and hardware implementation. As the size of the design space prohibits an exhaustive search, system architects have so far solved this issue in a rather pragmatic fashion by using an evolutionary design approach. Here, a base platform is enhanced and tailored for a particular application and market segment. Together with the extensive use of IP components with standardized interface, high reuse, and design efficiency along with short time-to-market can be achieved.

Because of the upcoming software challenges for MPSoC platforms, research and industry have developed so-called virtual platforms that contain all essential building blocks of embedded systems including processor cores and peripheral devices. The main target of virtual platforms is to support software development and platform analysis. Although recent research has increased the simulation speed and the modeling efficiency, the fundamental technique is still based on the principle of instruction set simulation. This inherently shows the same problems for early design stages as previously discussed in Sect. 3.1.4. Among them is the need for a fixed instruction set architecture along with a compiler and software. As neither one is typically fixed at particular early design stages and software development effort can be significant, a high risk exists that wrong design decisions only become visible after the implementation is finished and the effort is spent. To avoid such false decisions a clear and sophisticated methodology is needed that addresses early design phases. In the context of this chapter the following methodical aspects have been identified that should be covered by an *early* design space exploration methodology.

- *Aspects for efficient design space exploration*
  - *Guide system architects in their design decisions.* The size of the design space prohibits an exhaustive search and the simple application of theoretical exploration strategies. Therefore, practical relevant approaches address this issue according to the workbench idea, assisting system architects in their design decisions.
  - *Unbiased selection of design options.* Today's dominant design principles of platform- and component-based design [32, 33] induce the problem that designers tend to inspect design decisions with a certain bias [42]. For example, assuming a particular IP component exists, system architects tend to incorporate it into the system design without further questioning other implementations that eventually lead to an improved design. Therefore, a future design approach should *virtually* start with a blank sheet of paper and only consider these IP components as one option among others.
- *Aspects for efficient evaluation of a single design point*
  - *Support of arbitrary design entries.* A wide variety of entry points into the design space exploration have to be covered. For example, starting from scratch all design options, whether software or hardware, can be considered. In contrast, the reuse of a complete hardware platform restricts the possible design options to the software ones.
  - *Evaluation on multiple abstraction levels.* Induced by the different design entry points and stages, multiple abstraction levels need to coexist. At the start of the development typically only a rough idea of application, architecture, and mapping exists. Therefore, system architects utilize more abstract techniques based on estimates at such level, while in final design steps measurements based on fine-grained simulation techniques like instruction set simulation are adopted.

- *Smooth transition and back annotation.* An efficient design flow can only be guaranteed when a smooth transition from a high to a low abstraction level and backward is incorporated. In addition, different abstraction levels should be investigated jointly to achieve the best results. This demands incorporating investigations on low-level detail into the high-level models to capture these effects in the more abstract level.
- *Simple and fast modeling of a design point.* The commonly agreed orthogonalization of concerns [148] and the separation of interfaces and behavior [222] is key to efficiently evaluate particular design points.
- *Incorporation of all common software and hardware concepts.* The usage of multiple processor cores in today's and future platforms introduces software and hardware concepts that have to be considered for the precise characterization of a specific design point. This incorporates process concurrency on single and/or multiple processor cores including operating systems (OSs). Interprocess communication (IPC) might affect the performance heavily and hence has to be considered for evaluation as well.
- *Validation and verification.* Ensuring that all the characteristics are met in terms of performance and costs requires a detailed validation process. Moreover, this plays an important role in the final steps of the design verification to guarantee that the implementation operates according to its specifications. This has to be applied between different abstraction levels where one result serves as the reference for the following abstraction level.

These challenges have been addressed in various methodologies and tools targeting MPSoC design space exploration at particularly *early* design stages. The following chapter discusses these environments and methodologies.

Chapter 4
Related Work

This chapter reviews research activities relevant to the emerging challenges for design space exploration, in particular at *early* design stages. In the previous chapter the underlying techniques and ESL design have already been discussed. Hence, this chapter focuses on recent approaches operating on a higher abstraction level to overcome the limitations of virtual platforms and instruction set simulation.

During the earlier discussion three major aspects to be covered by design space exploration have been identified. These are given by the *evaluation of a single design point*, the *search strategy among the many design options*, and the *path from specification to the final implementation*.

In practice, several realizations of design space exploration frameworks exist, each having its own advantages and disadvantages. A rough classification separates the three main techniques of *simulation*-based, *analytical*, as well as *joint analytical and simulation*-based approaches. However, prior experiences from the domain of early HW/SW codesign helped to formalize the fundamental paradigms of orthogonalization of concerns [156] and the general Y-chart [103, 223] to be the principle for the most common frameworks.

Based on the prior classification, this chapter highlights the frameworks relevant for early design steps.

4.1 Simulation-Based Approaches

Simulation techniques commonly evaluate a single design point by tracing the execution characteristic during simulation runs. A brief introduction of various approaches is given next, while a detailed discussion can be found in [42, 77, 224].

The three main aspects of design space exploration are hardware architecture, application, and application-to-architecture mapping, commonly referred to as temporal and spatial task mapping. Early design space exploration frameworks must use higher abstraction levels as the issues discussed in Chap. 3 prohibit the use of instruction set simulation.

Typically, approaches provide an abstracted simulation model for the anticipated processing element within the final hardware architecture. Besides the pure

T. Kempf et al., *Multiprocessor Systems on Chip: Design Space Exploration*,
DOI 10.1007/978-1-4419-8153-0_4, © Springer Science+Business Media, LLC 2011

hardware aspect, software developers are supported by efficient programming models to simplify the development process and the inevitable complex application-to-architecture mapping. In general, such programming models have their genesis in the computer science domain based on process networks [225], in particular KPNs [19] and later component-based software engineering (CBSE) [26].

Apart from these fundamental characteristics, each approach targeting the complex field of design space exploration on the system-level and at particular early design stages, has its own underlying concepts and mechanisms. Hence, the following brief introduction tries to cover the most relevant simulation-based techniques.

CASSE is the acronym for CAmellia System-on-chip Simulation Environment [226], developed by Philips, NXP and the University of Las Palmas. The intended design process starts with an application model based on a KPN description. A simple and fast mechanism is provided to assemble the hardware architecture based on the main components given by processing, communication, and storage elements. Thanks to the Y-chart paradigm, the mapping step is reduced to a task to processing element assignment, whereas communication is strictly handled by the task transaction level (TTL) interface [227]. The technology has been successfully applied to multimedia applications [228]. In addition, a path to the final hardware implementation has been envisioned [229].

The *DAEDALUS* [129] project and design space exploration framework unifies the independent design methodologies of Sesame [125], ESPAM [230], and KPN-gen [225]. The resulting environment allows system-level exploration based on the high-level Sesame tool. Here applications are described as KPNs similar to CASSE. The KPNgen tool based on the technique of Compaan [231] offers support in the identification and separation of processes, finally leading to a KPN based application description. Both tools, Sesame and ESPAM, operate on IP components and their assembly for platform design. While the Sesame tool targets the domain of ESL design, ESPAM assembles components at the RTL level to bridge the gap to FPGA based prototyping.

The goal of the *ROSES* [141] framework and its extensions [232] is to increase IP component reuse and enhance software and hardware design. Following the paradigm of component-based platform design, it provides tools and mechanisms to easily assemble simulation models at high abstraction level, but also on RTL. Thanks to the technique of interface generation, the assembly task is significantly shortened. Besides the hardware modeling, a component-based software design can generate configuration properties of operating systems [233] and hardware dependent software (HdS) [141].

Performance modeling as proposed by Aldis is based on lessons learned from TI's OMAP-2 platform. It serves as an environment for a performance modeling, which helps to identify arbitrary execution characteristics. By exploiting the obtained results the final System-on-Chip design process is reduced to a straightforward integration. The model is composed of IP components that provide different accuracy levels, ranging from fully cycle-accurate models to generic traffic generators. Typically, peripherals, memories, and communication architectures are modeled fully cycle-accurate, whereas the framework provides three modeling styles for processor cores based on stochastic, trace-driven, and instruction set simulation.

Investigating *result oriented modeling*, Schirner et al. [234] propose a simulation model operating on timing annotations. It includes features to consider the influence of real-time operating systems and other low-level hardware effects, like interrupts. In addition, the software design can make use of a process network-based approach, which allows hardware dependent software (HdS) generation [235].

SystemCoDesigner [236], developed by Haubelt et al., focuses on the domain of DSP and offers the possibility to reuse multiple models of computation, e.g., synchronous dataflow (SDF) [20] graphs or KPN. For abstract system simulation, Virtual Processing Components (VPC) are assembled and allow a simulation-based system analysis. Finally, the framework provides an automatic FPGA back-end for fast prototyping based on the Embedded Development Kit (EDK) of Xilinx [237].

In *CoFluent Studio* [238] the development is performed on an abstract SystemC level, based on a message-passing principle. Here, early performance considerations are focused and can be efficiently analyzed. The tool suite includes a graphical design-entry tool and simulation framework.

Intentionally developed for the *Eclipse* [172] platform, Philips Research envisioned a framework that supports efficient design space exploration using KPNs. Addressing the specific domain of computationally intensive multimedia applications, the application-to-architecture mapping is supported by the explicit use of highly efficient communication architectures.

The design environment proposed by Herrera et al. [239] allows modeling applications based on process networks like SDF graphs and KPNs. Based on an abstract simulation model, design decisions can be analyzed including effects of real-time operating systems [240].

On the foundation of *ARTS* [134], the *UPPAAL* [241] framework has been designed. The solid mathematical definition of timed automata for task modeling allows efficient design space exploration with particular focus on scheduling issues within multi- and many-core platforms. To speed-up the design process only performance modeling is supported, while the functionality is generally neglected.

Apart from the various framework manifestations and techniques, two commonly utilized properties can be identified when comparing the different approaches. First, significant effort has been spent in the definition and development of *efficient programming models*. The approaches taken are based on *process networks* besides traditional C/C++ and Assembly based software development. The second key aspect is the development of an *abstract simulation model* enabling simulation on a higher abstraction level than instruction set simulation. Several techniques are based on *annotation* of the execution characteristic.

4.2 Analytical Approaches

As introduced within Sect. 3.1.2, formal and analytical approaches commonly operate on rather coarse-grained estimates of the underlying platform and implementation. This makes them applicable right from the start of the design cycle, but restricts evaluation mostly to WCET behavior and determinism must be assumed. Despite these restrictions, analytical approaches can help to significantly speed-up the exploration process, hence the most relevant techniques shall be sketched in the following.

From queuing theory of computer networks [188] and linear system theory [189] the technique of Network Calculus [190] has emerged operating on *event streams*.

The *Real-Time Calculus* [191] developed at ETH Zürich represents the application as arrival curves and the platform resources are captured as service curves. Based on a given application-to-architecture mapping, both are combined allowing the evaluation of the performance characteristic and resource utilization. This performance analysis has been successfully applied to the domain of network processors [192–194] and more general embedded systems [195].

The *SymTA/S* approach [197] operates on workload models [196] to allow the symbolic calculation of the execution characteristic. This approach is similar to the above-introduced real-time calculus. Hence, both have already been successfully combined under the SymTA/S framework [242]. Furthermore, this approach has been commercialized by a start-up called Symtavision. For a detailed discussion of this and the network calculus approach [198–200] should be consulted.

Despite the promising results of these approaches several pitfalls and accuracy issues have appeared with the currently available tools discussed in [243]. The introduced results show that these analysis methods encounter inaccuracies especially in the presence of control and data dependencies in the application task execution.

Apart from the class of frameworks operating on arrival curves or derivatives, other formal analysis frameworks [202] make use of *conditional process graphs* [203, 204]. Within these frameworks, the application tasks are mapped onto a specific hardware architecture thus allowing computation of the performance characteristic. The major goal of this work is to analyze and optimize scheduling effects within heterogeneous systems including application and hardware-architecture effects.

Another graph-based approach follows the principle of *SDF* graphs [20]. This solid mathematical foundation has been utilized for scheduling analysis with respect to single- and multiprocessors scenarios [86]. Both use cases have been extensively studied and have shown promising results in the evaluation of a single design point, e.g., in [205].

Timed automata [244] have been extensively studied and applied to analyze the scheduling of event-driven systems. Based on the formal definition of UPPAAL [241] the analysis has been applied successfully to in-car radio navigation system [245].

In general, it is observed that analytical approaches commonly operate separate from ESL design methodologies utilized today. Naturally, this leads to a significantly increased design effort as the analytical models and later the real implementation must be developed. In addition, a smooth migration from analytical- to simulation-based approaches is typically not available.

4.3 Joint Analytical and Simulation-Based Approaches

To strengthen either analytical or simulation-based approaches, a few approaches have been developed combining both technologies. However, use cases are mostly limited to special design issues, like the inspection of memory and cache behavior.

Trace-based performance analysis has been used extensively to reduce the necessary simulation time by doing costly performance measurements, e.g., of caches, once and then reusing the traces in the future. This allows inspection of various memory and cache architectures by utilizing these traces instead of doing time-intensive simulations repeatedly. The capabilities of this technique have been demonstrated by several case studies [207, 208].

Addressing communication architectures, Lahiri et al. [210] have extended the technique to *trace-driven analysis*. Similar to the evaluation of memories and caches based on once measured traces, the concept includes investigations of communication architectures. This allows the computation of performance analysis and resource utilization of architectures ranging from simple to complex Networks-on-Chips (NoCs). Further enhancements allow investigation of the impact of control operations [213] and resource contentions [214].

Another concept is based on the principle of *initial calibration* [215]. In a first step a large set of different benchmarks is measured which tries to cover the complete design space. In the following step, the attained results are used to estimate performance and power figures [216].

The earlier-mentioned approaches follow the paradigm to first measure the trace of a particular execution characteristic, which is then utilized to optimize the targeted design objective. Contrary to these, the approach proposed by Künzli et al. couples the real-time calculus with a SystemC simulation environment [217]. The major contribution is a converter outputting SystemC events from a given input described as an arrival curve and vice versa. These converters allow interfacing of formally described system parts with ones developed in SystemC.

In general, joint analytical and simulation-based approaches operate in two steps. In the first design phase the execution characteristic of one or a large set of benchmarks reflecting the common application characteristics is measured. In a second step the obtained information, commonly based on traces, is incorporated to evaluate the optimized system-level design analytically.

4.4 Summary

Throughout this chapter a large variety of techniques and methodologies for *early* design space exploration has been introduced. Because of the encountered limitations of virtual platforms and instruction set simulation, envisioned frameworks aim at higher abstraction levels to achieve higher modeling efficiency and to be applicable right from the start of the design process.

To address these challenges recent research activities have given birth to several approaches. However, analytical, simulation-based, or joint approaches are not able to cover all the specific demands. The main proposition of simulation-based approaches has been the development of simulation models on higher abstraction levels than instruction set simulation. Additionally, challenges in software development, in particular multiprocessor issues, have been targeted by programming models typically based on process networks. Pure analytical design space exploration methodologies operate on fairly abstract techniques leaving a huge gap between the proposed and the existing ESL design techniques. Joint approaches typically target only small parts of the overall design.

The lack of convenient modeling frameworks combining the set of different approaches in a unified environment hinders the practical use of the proposed methodologies. Accordingly, these environments commonly allow the inspection of one or the other objective, but are separated from the other existing exploration techniques. Hence, the design effort has to be invested twice, which most often results in abandoning the proposed approach.

In this book, a design space exploration framework is proposed that smoothly integrates on top of well-known ESL design techniques. The major contribution is two abstraction layers above instruction set simulation. The first introduced level is based on an abstract simulation. On top of this, a formal-analysis layer has been added with the purpose of quickly identifying corner cases and implementation candidates. In addition to these abstraction levels, the framework provides a unique mechanism to smoothly migrate from one level of abstraction to another, including virtual platforms. This opens the door to the final implementation by reusing common low level design techniques. The next chapter gives a brief introduction, followed by a detailed discussion in the rest of the book.

Chapter 5
Methodology

After the introduction of design space exploration and the discussion of related work, this chapter focuses on the methodology of the proposed design space exploration framework. As the design of software and hardware, along with the inherent question of temporal and spatial task mapping, is unfortunately not a simple query that can be answered with a simple yes or no answer, an iterative methodology is mandatory. Necessarily, system architects require a versatile framework which allows for simple and quick evaluation of arbitrary design decisions.

First, this chapter introduces the overall design process including an iterative refinement and evaluation flow. Within this chapter the different design steps will be briefly highlighted. In Chap. 6 a detailed discussion of the analytical implementation model is presented, whereas an in-depth discussion of the abstract simulation-based model is given in Chap. 7.

5.1 Iterative Design Process

Finding an optimal MPSoC solution for a particular application is virtually impossible because of the large design space defined by the many design options. Additionally, these various designs cannot all be evaluated. Therefore, design space exploration has to focus on the evaluation of only promising design points defined by the application, architecture, and application-to-architecture mapping. Besides this pure evaluation aspect, the framework needs to efficiently guide system architects through the immense design space to identify the best possible solution. Fundamental aspects during this exploration step are the following ones.

- *Application aspect.* Commonly given as a written specification document or defined by mathematical equations, the application needs to be partitioned into tasks. This partition step includes identification and extraction of parallelism within the application. Definitely, the selection of tasks has to obey the underlying hardware characteristics. Therefore, a tight relationship exists between the application and hardware aspect discussed later.
- *Hardware aspect.* Selection of suitable hardware components is key since high performance demands have to be kept under tight energy constraints

T. Kempf et al., *Multiprocessor Systems on Chip: Design Space Exploration*,
DOI 10.1007/978-1-4419-8153-0_5, © Springer Science+Business Media, LLC 2011

especially for battery powered devices. Hence, selecting suitable components requires a sophisticated identification process considering coarse- and fine-grained hardware details. Coarse-grained considerations define the selection of either programmable processing elements like GPPs, DSPs, and ASIPs, or nonprogrammable components such as specially tailored ASICs to perform a dedicated task. More fine-grained selections operate on much lower hardware details such as the definition of instructions to speed-up execution of a specific task. Such considerations are typically referred to as Instruction Set Extension (ISE) like the ones in [48, 56, 246, 247]. Other frameworks allow modifying and defining the complete instruction set architecture of processor cores, e.g., [49, 248].

- *Application-to-architecture mapping.* Temporal and spatial mapping of a given application to a particular architecture creates serious challenges. In this context temporal mapping defines the execution of tasks mapped onto a single processing element and the spatial mapping denotes the distribution of tasks over the different platform components. The selection of a mapping strongly influences the other two issues of design space exploration, namely the identification of hardware processing elements and the selection of suitable software implementation characteristics.
- *Selection of software implementation characteristic.* Choosing the right software implementation in terms of development effort and costs vs. the achievable performance requires deep investigations. In general, the impact on the system performance is significant and performance gains can mostly be traded against energy efficiency by, for example, scaling the clock frequency. From the perspective of development effort, high-level programming languages are preferable. Such high-level languages provide inherent constructs that allow for a fast and structured software implementation. However, usage of these languages typically results in overhead within the compiled executable. Especially, this applies to highly irregular hardware structures, e.g., ASIPs and to some extent to DSPs. Therefore, such architectures are typically programmed with low-level software based on Assembly programming or the use of inline Assembly and compiler-known functions [249]. These different implementation options make trade-off decisions necessary and developers have to decide whether to invest significant effort in low-level programming or to accept the induced overhead by high-level languages. To take an optimal or at least near-optimal design decision a thorough exploration phase is required.

Besides the large variety of possible options for MPSoC development, the design entry can differ significantly as already shown in Fig. 3.3.

- The design cycle might start from a blank sheet of paper
- A major redesign of an existing platform
- An enhancement of an existing platform or
- A complete reuse of an existing platform

Certainly, the entry point into the design process has a major impact on the available implementation options and the design approach. For example, starting from scratch, the system architect can apply any kind of coarse- and fine-grained system optimizations to achieve the overall requirements. Contrary to this, reuse of an existing platform restricts the design space to software modifications, which clearly limits the applicable optimizations during the design of a system.

All methods used for design space exploration, whether analytical- or simulation-based, rely on the characterization of the hardware and software. To obtain reliable analysis results a sufficiently precise input characterization of both is needed. This inherently demands that during design time, characterizations are constantly updated and improved in their precision till the final implementation is available. As previously discussed, such characterization depends deeply on several parameters like the implementation characteristic in terms of the coding style of software, the modeling style of the hardware platform including processing elements, and the communication and memory architecture. To cover all these issues a design process is proposed based on an iterative refinement loop as highlighted in Fig. 5.1.

The performance parameters serve as an optimization criterion for the later design process. In the area of wireless communication these parameters usually relate

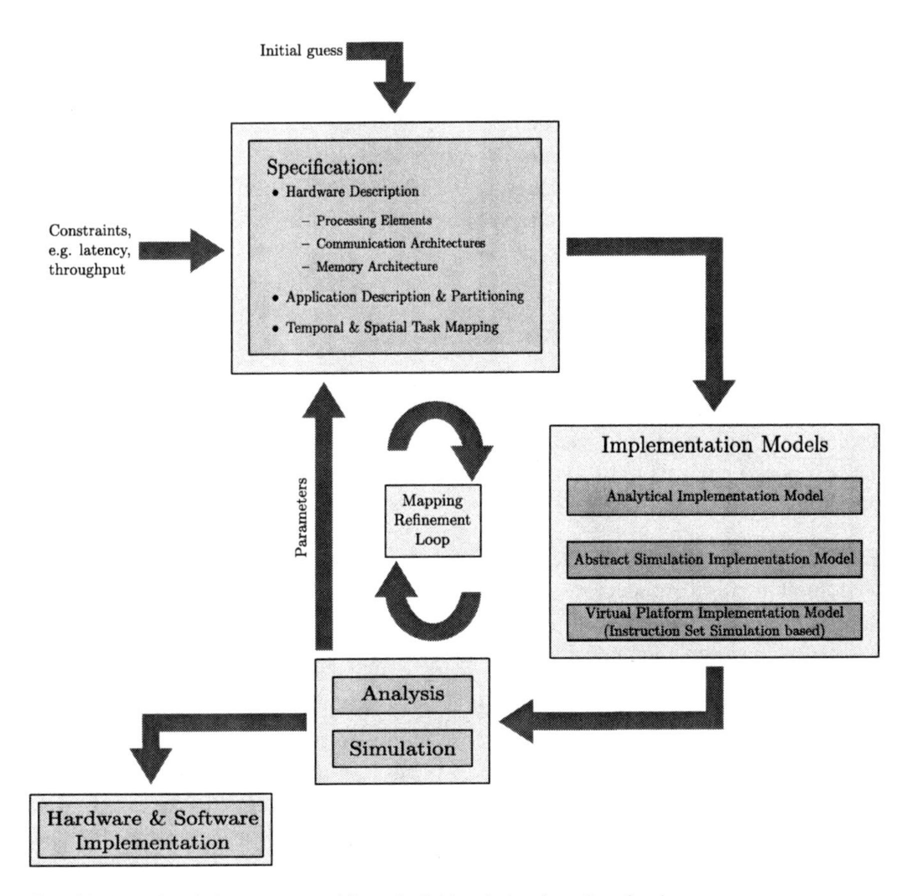

Fig. 5.1 Iterative design process with analysis/simulation-based evaluation

to latency and throughput requirements. Other typical constraints relate to memory size and buffering requirements. The design loop is entered with an initial guess at a suitable hardware platform and a temporal and spatial task mapping. In a first design iteration, an implementation model based on estimated parameters is composed. Based on this model, system architects can iterate over different design decisions and evaluate their characteristics by the proposed mathematical analysis discussed in Chap. 6. Having identified one or multiple implementation candidates, the implementation model can be refined to the next abstraction level. Moving from the formal analysis to an abstract simulation, the framework enables evaluating the implementation candidates in a more fine-grained manner. The underlying abstract simulation-based technique is discussed in-depth in Chap. 7. The key component for abstract simulation-based design space exploration is the generic Virtual Processing Unit (VPU) [250]. At early design stages it allows reuse of the analytical models. Later, model refinement is carried out till an implementation model based on instruction set simulation is available. This can serve as basis for lower level implementations like Register-Transfer-Level (RTL) and later. Unfortunately, the design process is mostly not that straightforward, thus at each stage the methodology offers a back annotation of the necessary information. This enables an iterative design process which interweaves analytical-based design space exploration with simulation-based techniques. In addition, the complete framework allows for a continuous refinement and exploration flow from pure mathematical analysis to the final implementation. Before discussing the analytical- and simulation-based design space exploration in detail, this chapter highlights the principle of both.

5.2 Analytical Implementation Model

Applications in the domain of SDRs, wireless communication and also multimedia applications possess tight constraints for *critical paths (CPs)* and *feedback loops*. The proposed analysis aims to determine the three main issues [251] at an earliest possible design stage.

1. Identification of a suitable hardware architecture including processing elements, communication architecture, and memories.
2. Partitioning of the application into tasks.
3. Finding a suitable temporal and spatial task mapping.

The proposed mathematical analysis is based on the concept of a workbench, thus *no* automatic mapping and hardware architecture exploration is carried out. In contrast, the workbench allows easy evaluation of a specified design point. This helps to analyze whether a given constraint has been accomplished or not. In case of failure, the analysis guides designers to where modifications should be applied, whether to the hardware platform, the application, or the mapping. Additionally, the workbench supports designers in the implementation order of tasks such that performance-critical parts of the design are addressed first.

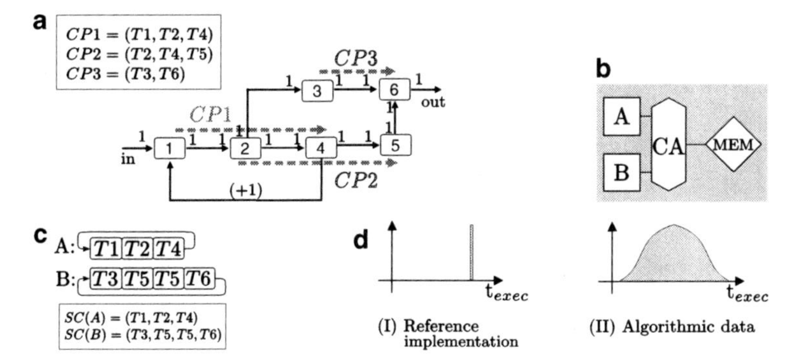

Fig. 5.2 Exemplary analysis components. (**a**) Task graph and critical paths. (**b**) Hardware architecture. (**c**) Temporal & spatial task mapping. (**d**) Task characteristic examples $X(Task, PE)$

Applications within the design space of wireless communications can be adequately described as one or multiple SDF [20] task graphs. For example, each mode of a standard can be consistently described as one task graph, while the combination of multiple disconnected graphs describe the complete standard.

Before going into a detailed discussion of the analytical implementation model, an overview will be given in the following part of this section. Figure 5.2 shows a simplified example to highlight the analysis. The general application decomposed into multiple tasks (Fig. 5.2a) is mapped temporally and spatially (Fig. 5.2c) on the depicted HW architecture (Fig. 5.2b). For this particular case the workbench allows evaluation of the latency and throughput of the critical paths (CPs) (Fig. 5.2a). This evaluation needs sufficiently precise characterization of the processing and communication behavior of each task on its mapped processing element (Fig. 5.2d). Unfortunately, these are not known, or bound with a particular uncertainty at the addressed early design stage. Thus, an iterative design loop is mandatory to validate the system at each design step with the best available knowledge.

The task's execution and communication characteristic is conceived in a *random variable*. The uncertainty of these characteristics is tightly coupled to the designer's knowledge of the task, ranging from complete knowledge of the execution, e.g., when a reference implementation is available, to a merely algorithmic level where only rough estimates of the required operations exist. These characteristics can be described as a *probability density function (pdf)* given, e.g., in the first case by a dirac delta function (Fig. 5.2d.I), whereas it might be given in a second case as a Gaussian-like distribution with a high variance (Fig. 5.2d.II). Please note that the proposed analysis operates on these estimates, so that better ones obtain results which better match the real final implementation. This implies a degree of uncertainty at the

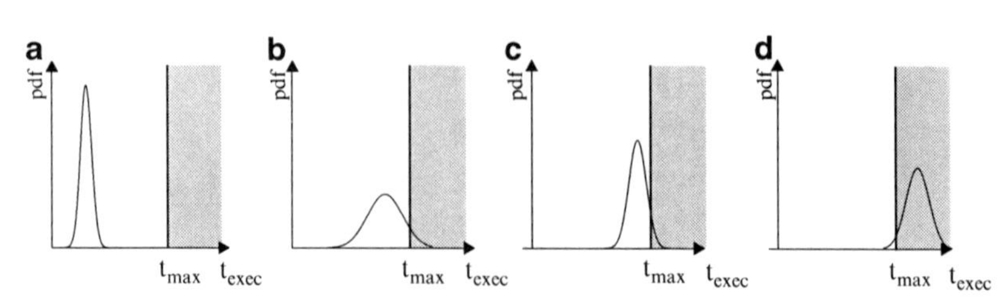

Fig. 5.3 Exemplary analysis results for latency constraints. (**a**) Likely feasible. (**b**) Uncertainty dominated. (**c**) Expected value dominated. (**d**) Unlikely feasible

beginning of the design phase. During the design process, measurements, for instance gained by simulation, can provide more specific implementation knowledge which can help to reduce the uncertainty.

As the analysis is based on estimates, the result cannot just be given by a boolean decision, rather it is based on a likelihood whether the analyzed system is feasible or not. Therefore, the result for each critical path is a random variable determined by its probability density function (pdf). Figure 5.3 depicts possible results.

- *Feasible (Fig. 5.3a)*: The highlighted random variable is most likely to keep the threshold. Depending on the margin, developers might consider modifications of the hardware and/or task mapping because such systems tend to be overdesigned.
- *Uncertainty dominated (Fig. 5.3b)*: The high margin between threshold and expected value makes it likely that in spite of the failure probability, which might be caused by imprecise estimates, the system should work. In such cases developers should focus on the implementation of tasks with a high uncertainty, respectively tasks with imprecise knowledge, first. After implementation more precise values can be determined and analysis should be re-run to verify that the addressed implementation still meets the constraints.
- *Expected value dominated (Fig. 5.3c)*: Here the probability density function shows only a minor variance, but only a small margin between the expected value and the threshold exists. Such systems tend to fail because of unexpected behavior. Therefore, system architects should inspect carefully if the envisioned implementation has to be improved or if the system really has to work at its limits to achieve this requirement.
- *Unlikely feasible (Fig. 5.3d)*: A high probability exists that the given constraint cannot be satisfied, therefore system architects have to re-consider either the addressed hardware or task mapping.

As the analysis results rely on the given input parameters, the overall uncertainty of the analysis results is dominated by the uncertainty of these parameters. At early design stage these input parameters are typically estimates based on expert knowledge. Therefore, these estimates contain a certain degree of imprecision, which is reduced during the design process by replacing the estimates with fine-grained implementation knowledge, e.g., obtained by simulation traces. Inherently, this leads

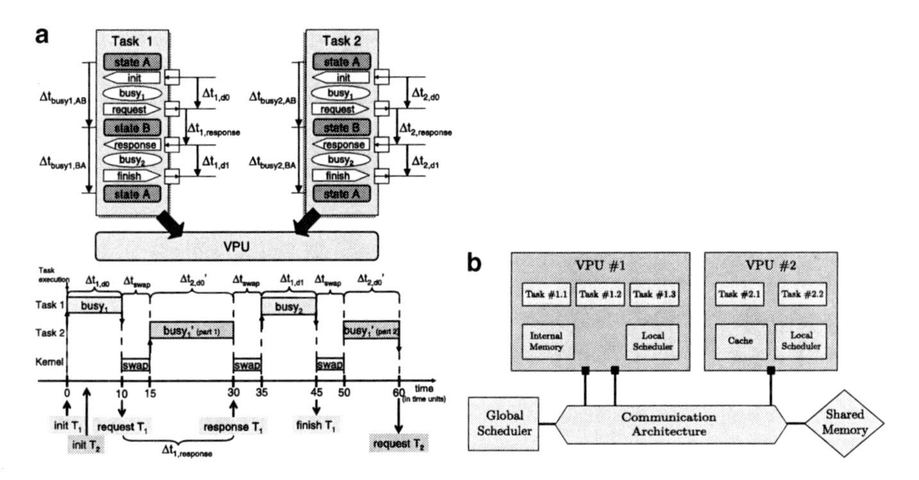

Fig. 5.4 Principle and usecase of the Virtual Processing Unit (VPU). (**a**) VPU Performance Model. (**b**) System-level design including VPUs

to the refinement flow sketched in the overview (Fig. 5.1). Apart from this, the back annotation keeps higher abstraction levels consistent, supporting arbitrary switching from high- to low-level of abstraction and vice versa. After the identification of suitable implementation candidates, the analytical implementation model is refined to an abstract simulation model and finally to an instruction set simulator based implementation model. In the following section the principle of the abstract simulation model is outlined, whereas a detailed discussion follows in Chap. 7 (Fig. 5.4).

5.3 Abstract Simulation Implementation Model

The implementation model for abstract simulation utilizes the technique of timing annotation similar to the ones described in [252–254]. In the following, the technique of timing annotation for design space exploration is discussed based on the framework introduced in [250]. Central component of this framework is the so-called VPU, which can be configured to imitate the behavior of arbitrary processor cores within a system-level simulation. The simulation technique is based on the principle of annotation, which allows the modeling of software execution exclusively on the basis of execution characteristic annotations with no functionality. In a later refinement stage the functionality is iteratively included till the final implementation is available. The annotation concept and principle will be sketched exemplarily on the basis of the VPU later.

The example depicted in Fig. 5.3 illustrates the annotation and VPU mapping mechanism. The upper part of the figure illustrates two tasks with their individual execution characterstics which are mapped to a single VPU instance. The lower part of Fig. 5.3 shows the resulting behavior of the VPU according to an assumed scenario, which will be discussed in the following:

First task 1 is activated by the external *init* T_1 event and executes the first portion of the task. The simulated execution time directly corresponds to the annotated time $\Delta t_{1,d0}$. Before entering state B task 1 initiates an external data transfer request. Although T_1 waits for the response to this request, task 2 can execute. First a task swap, e.g., initiated by an Operating System (OS), is performed which requires 5 time units for the given example so that task 2 can start execution after 15 time units. The VPU takes care that this swapping time is taken into account and shields the tasks from external events. In the given scenario execution of task 2 requires more time than the response of the data transfer of task 1. Assuming task 1 has higher priority than task 2, a task preemption occurs and task 2 cannot be resumed before the second portion of task 1 has completed its functionality. The request generated by task 2 is delayed by the VPU till the correct point in time is due. Thus, from the perspective of external system components, the external events are visible at the corresponding time of concurrent task execution.

The VPU concept allows modeling of processor cores supporting concurrent task execution by, e.g., operating systems or hardware multithreading. For system-level simulation, multiple VPUs where each mimics a different processor core, can be assembled like shown in Fig. 5.3. This supports the evaluation of different design decisions in a quick and simple manner. Typical goals of system evaluations are:

- Identification of the number and type of processor cores.
- Identification of the connecting communication architecture and necessary storage elements.
- Identification of the application to architecture mapping.

One key issue while utilizing annotation-based simulation is how to obtain these execution characteristics. Especially at early design stages, no software implementation or merely nonoptimized functional implementation of the intended application exist. Because of this lack, identification of the execution characteristics for a particular application can be rather complex. Thus, efficient design space exploration requires an iterative design process, starting at high abstraction level with only rough estimates. Such estimates can be reused from the previously described analytical implementation model by utilizing the Time Retrieval Engine as later discussed in Sect. 7.6. In the subsequent refinement loop the estimates are continuously improved till the complete implementation is finally available. To allow such an iterative design process the VPU supports different levels of software modeling and annotation. The supported ones are illustrated in Fig. 5.5.

At the highest abstraction level, the execution characteristic is modeled based on statistical functions operating on random variables. The common application and architecture description of the framework efficiently guarantees the exchange of the characterizations between the analytical and abstract simulation-based implementation model. In the later stages where at least a rough understanding of the algorithm exists, developers can easily modify the annotations. First this can be based on complete tasks, whereas in later stages these annotation statements can be added within the software implementation. Please note that these abstraction levels, in addition to the VPU, support simulation without having any software

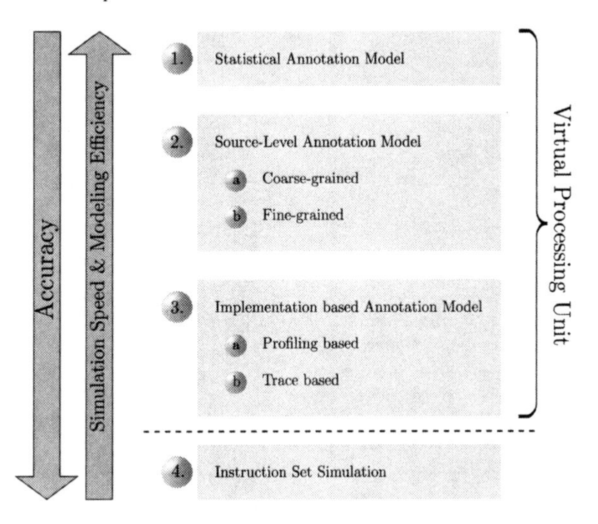

Fig. 5.5 Supported annotation models of the VPU

implementation at hand. More fine-grained annotations based on the μProfiler [255] or trace-based instrumentation, naturally require a software implementation. In case more detailed information about a task and its execution characteristic exists, higher abstraction levels can be skipped and the refinement loop can be entered according to the available knowledge.

Addressing applications from the domain of mobile terminals, a common description is conceived. Besides a traditional textual design entry (Sect. 7.5.1), the proposed framework includes a graphical design entry for an efficient and quick development process (Sect. 7.5.2). Especially, this design entry is well suited for applications from the domain of wireless communications and allows an application modeling based on task graphs. In addition, the temporal and spatial task mapping is simplified in the graphical environment to a drag and drop fashion (Fig. 5.6).

Based on the discussed principle of the abstract simulation model, system architects can evaluate arbitrary aspects of design decisions in a simulation-based environment. Results or encountered inaccuracies within the analytical implementation model can be improved by back annotating characteristics evaluated within the simulation environment. For example, contentions on the interconnect architectures can be one source for the need of such back annotations.

When designers are satisfied with their achieved results and would like to proceed further to the final implementation, they can utilize the framework's implementation link and refinement flow. This flow enables a continuous path from abstract simulation-based environment to fine-grained instruction set simulation (ISS). ISS-based simulation models are well-known and often used for software development as well as for hardware platform design and performance evaluation. Therefore, this technique is briefly introduced in the next section.

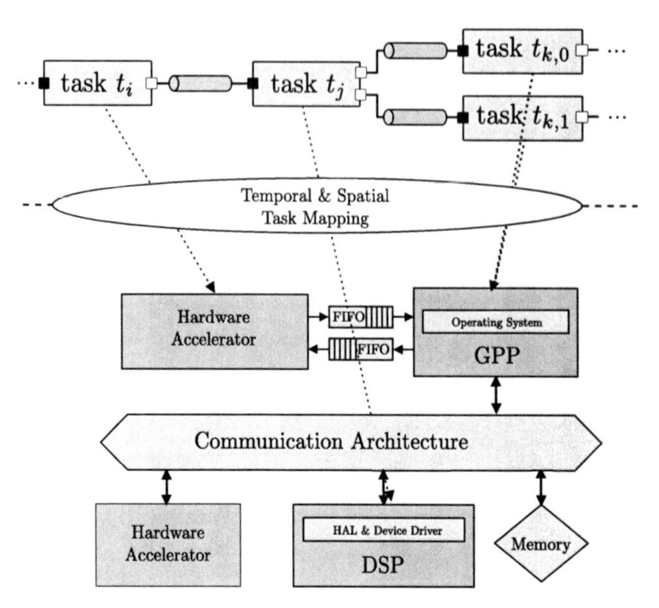

Fig. 5.6 Graphical design entry

5.4 ISS-Based Implementation Model

Following the refinement process of hardware and software including the annotation models, developers finally reach the highest achievable detail of the abstract simulation model. This level is characterized by a functionally correct software execution along with a detailed annotation of the execution characteristic. However, the underlying simulation model relies on the VPU technology mimicking the behavior of the real hardware. Quite naturally the next step moving toward the real hardware platform is to lower the abstraction by replacing the abstract simulation vehicle by an ISS. Additionally, software has to be cross-compiled for the targeted instruction set architecture, which inherently demands a cross-compiler.

Figure 5.7 highlights the basic refinement steps from the abstract simulation model based on the VPU technology to an ISS centric model. Since both technologies are based on the SystemC language, refinement of the hardware simulation model merely requires the replacement of the VPU with the underlying processor core, i.e., its instruction set simulator.[1] The key enabler for such replacement is the use of well-defined Transaction Level Modeling (TLM) 2.0 interfaces [153], with which the VPU is compliant.

Along with the modification of the hardware simulation model, major modifications apply also to the software part. The VPU technology executes the functionality

[1] ISSs do not necessarily have to be developed in SystemC language. However, most of the today prominent IP vendors of processor cores support integration of their proprietary ISSs into a SystemC environment by encapsulating it into a TLM-2 compatible simulation model.

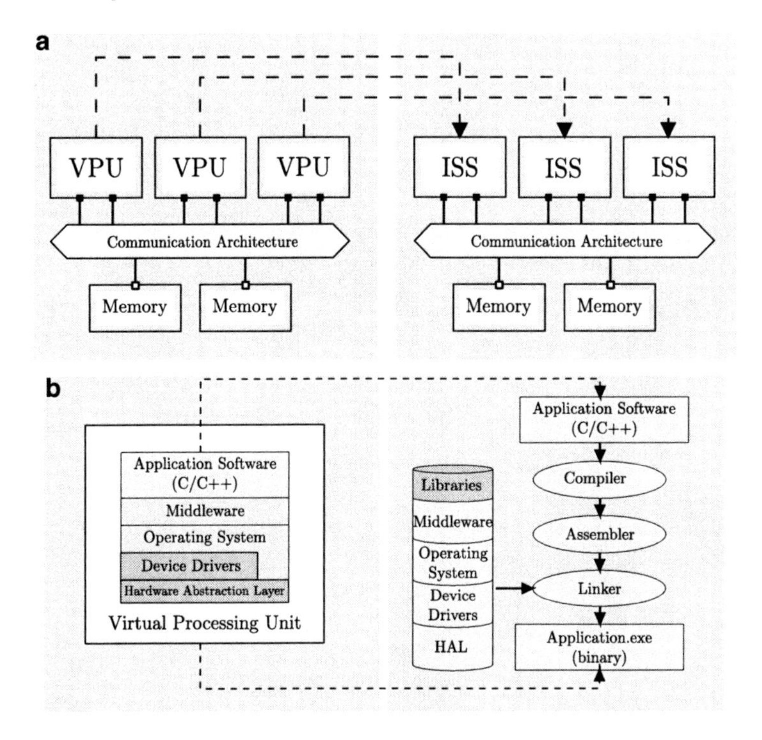

Fig. 5.7 (**a**) VPU to ISS refinement: hardware part. (**b**) VPU to ISS refinement: software part

of the software directly on the host processor. Execution characteristics are exclusively taken from the annotation statements to reflect the execution behavior of a particular processor core different from the host one. With the exchange of VPUs by ISSs, which imitate the exact behavior of a specific processor core, the software development has to necessarily follow the development flow of the targeted processor core. This inherently needs a cross-compilation of the developed software for the particular processor core. Certainly this requires the availability of several software tools and libraries such as a cross-compiler, an assembler and run-time libraries. Instead of using the abstract annotations, the ISS executes each instruction of the target executable step-by-step.

Lower level simulations like RTL and below are mandatory for later implementation of the hardware. However, for the proposed framework and design process, the well-known technique of instruction set simulation is taken as the lowest level of abstraction. Already past research has demonstrated paths to the final silicon implementation [129] or FPGA prototyping [256].

After the comprehensive introduction of the overall methodology, a more detailed discussion of the different parts of the framework follows in the next chapters.

Chapter 6
Analytical Implementation Model

After a brief recapitulation of the motivation and the fundamental problem, the discussion turns to the analytical implementation model.

6.1 Design Space Exploration as a Mathematical Problem

Design of embedded systems, especially wireless communication devices, has to deal with many highly complex issues (see Chap. 2). To design a cost- and energy-efficient system that meets the stringent constraints, early design space exploration is a must. The major objective of design space exploration is to support and guide system architects to take the right design decisions. Among others, the three key objectives are:

1. Partitioning of the application into a set of tasks.
2. Identification of a suitable hardware platform architecture.
3. Mapping the application, i.e., the partitioned tasks, onto the hardware architecture.

Figure 6.1 illustrates the evaluation problem of design space exploration and the proposed analysis. The well-known Y-chart principle [223] forms the foundation of the proposed analysis workbench. The main objectives that are subject to evaluation with respect to the given constraints determine the design parameters. Classification of these parameters separates the application, hardware architecture, and temporal and spatial task mapping. The analysis is structured as a workbench that allows evaluation and analysis of any given input. The obtained results support system architects to take the right design decisions.

Unfortunately, at early design stages mostly imprecise implementation knowledge and no final implementation exist. Therefore, the analysis needs to operate on imprecise input characteristics leading to uncertainties within the results. This prevents giving results in terms of boolean decisions, which decide whether a system works properly or not. In contrast, the retrieved results can only judge how likely the system might operate within a particular performance range at such early design stages.

T. Kempf et al., *Multiprocessor Systems on Chip: Design Space Exploration*,
DOI 10.1007/978-1-4419-8153-0_6,

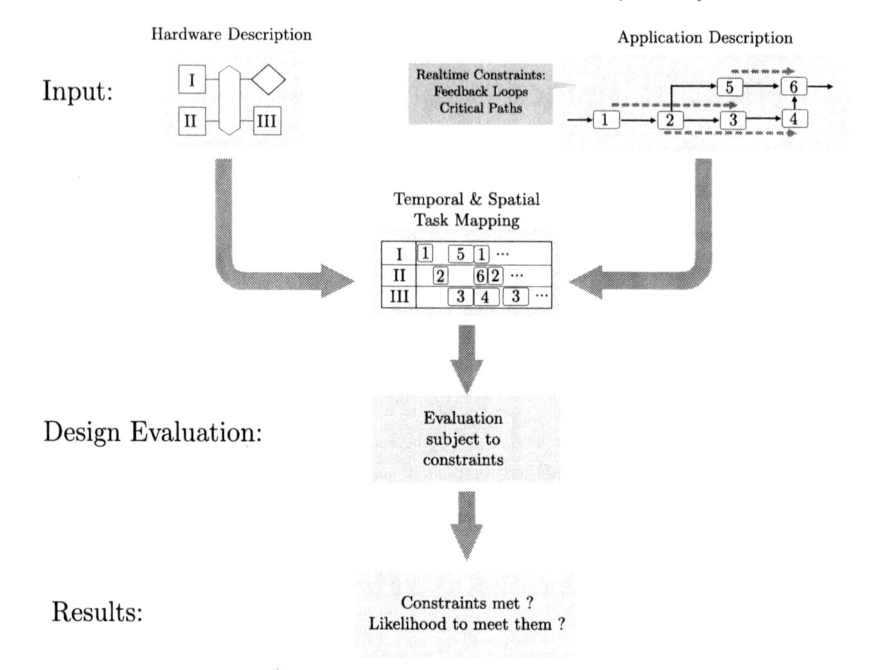

Fig. 6.1 Problem statement of design space exploration

The proposed workbench approach, as well as the uncertainty in the obtained results, makes an iterative design approach necessary. A typical design process starts with a fixed set of applications to be implemented on a device so that the requirements and constraints are met. Development of the hardware platform from scratch is only performed in rare cases, while the typical design starts with an initial guess at the platform or extends an existing base architecture to support the new applications. At early design phases it is essential to determine the possible implementation candidates which will fulfill the requirements. A well-structured identification prevents or at least minimizes the probability of time- and cost-intensive major re-designs in the subsequent implementation process. As multiple abstraction layers are mandatory, the analysis needs to be combined with later simulation-based approaches that are used to inspect the platform implementation in detail. Additionally, system architects must be able to directly start evaluation and exploration prior to cost- and time-intensive implementations. This definitely implies that the analysis operates on estimates because mostly no final implementation exists. The gained experiences and analysis results prevent system architects from taking false design decisions, hence saving cost and development time.

In the following, the proposed mathematical-based analysis that targets especially very early design phases is introduced.

6.1.1 Problem Statement and Elementary Definitions

The problem of design space exploration represents a multiobjective optimization problem. First reported by Vilfried Pareto, it is also referred to as *Pareto-optimization* [218].

Definition 6.1 (Multiobjective Optimization). Given a possible design represented as a decision vector $\underline{x} = (x_1, ..., x_k)^{\mathrm{T}}$ in the decision space $X \subseteq R^k, k > 1$, the function $f(\underline{x}) : X \rightarrow Y$ assigns each decision vector $\underline{x_i}$ of the decision space X a corresponding objective vector $\underline{y} = (y_1, ..., y_l)^{\mathrm{T}}$ in the objective space $Y \subseteq R^l, l \geq 1$ (Fig. 6.2). To identify the optimal decision vector $\underline{x}_{\mathrm{opt}}$ in the decision space X it is assumed without loss of generality that all objectives $y_1, ..., y_l$ are to be minimized

$$\min\{y_j\}\ \forall j \in \{1, ..., l\}.$$

It would be desirable to minimize all objectives at the same time, but due to contradicting objectives this is not possible in general. One example is the AT- or ATE-product in integrated circuit design as discussed in Sect. 3.2.

To perform well-balanced trade-off decisions *Pareto-dominance* guides designers to determine decision vectors best for at least one objective $\underline{x}_{\mathrm{dom}} \in X$.

Definition 6.2 (Pareto-dominance, Set, and Front). An objective vector $\underline{y}_1$ *dominates* ("is preferred to") another objective vector $\underline{y}_2$ ($\underline{y}_1 \succ \underline{y}_2$) if each component/objective of $\underline{y}_1$ is less or equal than the corresponding component/objective of $\underline{y}_2$ and at least one component/objective is strictly less. Therefore, a decision vector $\underline{x}_1$ is defined as *Pareto-optimal* or *-dominant*, if there is *no* decision vector $\underline{x}_2 \in X$ such that

$$\underline{y}_1, \underline{y}_2 \in Y,$$
$$\underline{y}_1 = f(\underline{x_1}) = (y_{1,1}, y_{1,2}, ..., y_{1,l}),$$

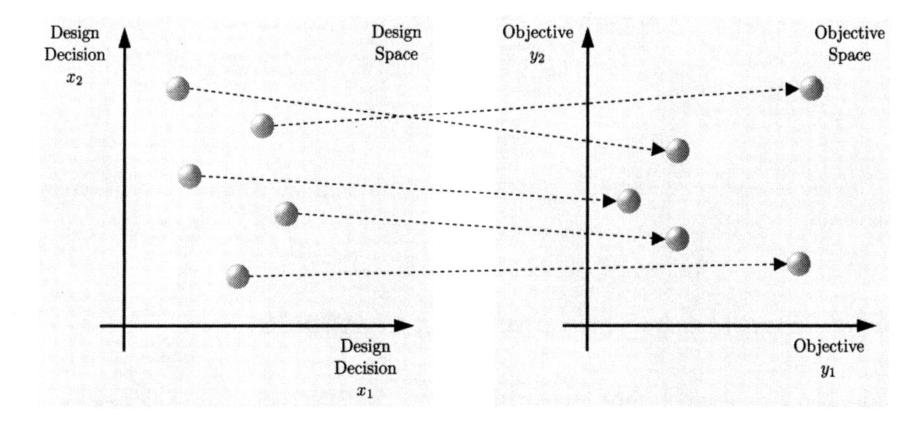

Fig. 6.2 Multiobjective optimization problem: decision and objective space

$$\underline{y}_2 = f(\underline{x}_1) = (y_{2,1}, y_{2,2}, \ldots, y_{2,l}),$$
$$\forall i \in \{1, \ldots, l\},\ y_{1,i} \leq y_{2,i}, \text{ and}$$
$$\exists i_0 \in \{1, \ldots, k\} \mid y_{1,i_0} < y_{2,i_0}.$$

The set of decision vectors that fulfill this condition are generally denoted as *Pareto-set* $X^* \subseteq X$ and the image of the set within the objective space is denoted as the *Pareto-front* $Y^* = f(X^*) \subseteq Y$.

To map the general multiobjective optimization problem to design space exploration, especially targeting early design phases, the required components of the decision and objective vector need to be specified. CbD [32] and the Y-chart principle [223] let the decision vector be abstracted into three dimensions defining the application, the hardware architecture and the temporal and spatial task mapping.

$$\underline{x} = (x_{\text{Appl}}, x_{\text{Arch}}, x_{\text{Map}})^{\text{T}}$$

Directly from the start of the design cycle, each component of the decision vector can be consistently addressed with such representation. In contrast, manipulating the objective vector consisting of latency, throughput, algorithmic performance, energy and area consumption requires typically intensive experiments.

$$\underline{y} = (y_{\text{latency}}, y_{\text{throughput}}, y_{\text{algo.perf.}}, y_{\text{energy}}, y_{\text{area}}, \ldots)^{\text{T}}$$

Latency and throughput objectives mostly decide whether the system is operational at all. These are referred as *hard* objectives. Other objectives such as energy consumption influence, the stand-by-time, and run-time of a device. This can clearly decide about the potential business success, but does not necessarily result in a complete system failure. Therefore, such objectives are further classified as *weak* objectives.

Especially at early design stages, objectives like energy and area consumption, can hardly be addressed. This sets the focus on the hard objectives like latency and throughput to guarantee a properly working system. Therefore, design space exploration defines the search for a *Pareto-optimal* solution $\underline{x}_{\text{DSE}} \in X$ so that the objective vector $\underline{y}_{\text{DSE}} = f(\underline{x}_{\text{DSE}}) \in Y$ fulfills the given objective constraints $\underline{c}$ in every single objective.

$$\underline{y}_{\text{DSE}} \leq \underline{c}$$

In the following, these three components along with the constraints are derived in detail.

6.1.2 *Input Analysis and Evaluation Constraints*

Based on the principle of the Y-chart, the mathematical design space exploration defines the application, the hardware architecture, along with the temporal and spatial task mapping as the fundamental inputs of the evaluation. In a following evaluation step, these are investigated subject to given requirements. After defining the input parameters, this section concludes with the evaluated constraints.

Application (x_{Appl})

As the analysis aims at evaluation of a complete device such as an SDR, the general assumption is that multiple applications are executed simultaneously on a single platform. From application perspective these communication standards are completely separate at least for the physical layer. However, the combined execution on a shared processing resource introduces dependencies between them. The following discussion separates the single application description from the joint description of multiple applications that is introduced later on. As assumed by most graph operations, the application task graph is transformed into a directed acyclic data flow graph [257] while preserving all contained information. This transformation follows the basic application description.

Definition 6.3 (Single Application Scenario). A single application such as physical layer processing can be consistently and efficiently managed as an SDF *[20] task graph* TG. Therefore, the initial task graph description of such application is defined as a directed graph:

$$
\begin{aligned}
TG = (T, D, r, \delta):& \\
& T \text{ is the set of tasks } \{t_1, ..., t_n\}, \\
& D \text{ is a set of ordered pairs } \{(t_i, t_j) : t_i, t_j \in T\}, \\
& r : D \rightarrow \mathbb{N} \times \mathbb{N} \text{ are the communication rates} \\
& \delta : D \rightarrow \mathbb{N} \text{ are the delay annotations}
\end{aligned}
$$

In addition, the graph is *consistent* meaning a valid schedule exists. Several methods and analysis techniques for identifying consistency can be found in [20].

The task graph is composed of tasks[1] T forming the nodes and edges D representing the data flow. The edges, respectively the data flow (t_i, t_j), have a static rate annotation r, that is a (x, y) ratio. Here x defines the number of tokens produced by task t_i and y is the number of tokens consumed by task t_j in each iteration. This demands that the iteration number of t_i and t_j must preserve consistency so that the task activations is given by:

$$\frac{N(t_i)}{N(t_j)} = \frac{y}{x}, \ \forall d_k = (t_i, t_j) \in D, \ r(d_k) = (x, y).$$

[1] For simplification, tasks consume all incoming data at the same time. Same applies for producing data on the outgoing edges. Therefore, modeling of data input and output at different times requires task splitting.

When the condition holds for every task within the graph and all tasks execute $N(t_i)$ times, all virtual buffers on every edge will have been filled and emptied completely. Otherwise buffer overflows are certain. This task graph description inherently captures all possible data flow information.

For further clarification Fig. 6.3 depicts three exemplary task graphs. According to the task graph definition, the example in Fig. 6.3c is a valid task graph while the two others represent illegal task graphs. The task graph in Fig. 6.3a violates the definition twice. First, an illegal loop exists among the tasks 3, 4 and 5. Second, there is a rate mismatch between the two paths from task 1 to 4. While the path[2] $(1,2,3,4)$ defines the rate as $2 \times N(1) = N(4)$ the direct path $(1,4)$ induces $N(1) = N(4)$. This rate mismatch would finally lead to a buffer overflow in the edge $(1,4)$.

Adding the delay annotation $(+1)$ to the loop, the second example (Fig. 6.3b) remains invalid as the rate mismatch still occurs. In the final example (Fig. 6.3c) the rate modification of the edge $(2,3)$ preserves consistency to the given task graph definition.

Definition 6.4 (Multiapplication Scenario). The previous definition of the single application scenario as a task graph does not exclude a multiapplication scenario. In other words j-applications each defined by a task graph TG_i, $i = 1,...,j$ can be summarized as a single application with the joint task graph TG_{joint} given by

$$TG_{\text{joint}} = TG_1 \cup ... \cup TG_j.$$

Figure 6.4 highlights such joint representation where two task graphs TG_0 and TG_1 are combined in a joint task graph TG_{joint}. However, this combined graph consists of two disconnected[3] graphs. Please note that control signals from upper protocol layers may connect both task graphs, which have to be modeled as well.

In the following, the generation of the *directed acyclic Data Flow Graph (DFG)* is discussed. Although all information remains preserved, a directed acyclic DAG structure is beneficial for later analysis and increases the freedom while mapping the application down to the hardware. The generation consists of two steps as illustrated within Fig. 6.5, namely

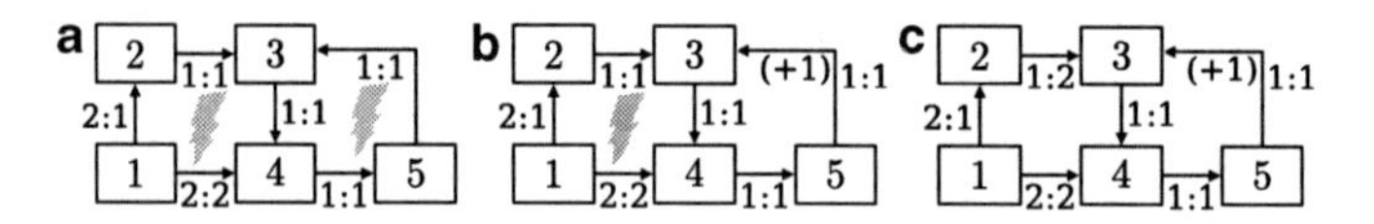

Fig. 6.3 Examples of valid and invalid application task graphs (TG). (**a**) Illegal task graph. (**b**) Inconsistent data rates. (**c**) Valid task graph

[2] Reference [257] defines: *A path is a simple graph whose vertices can be ordered so that two vertices are adjacent if and only if they are consecutive in the list.*

[3] Reference [257] defines: *A graph G is connected if it has a u, v-path whenever vertices u, v $\in V(G)$ (otherwise, G is disconnected).*

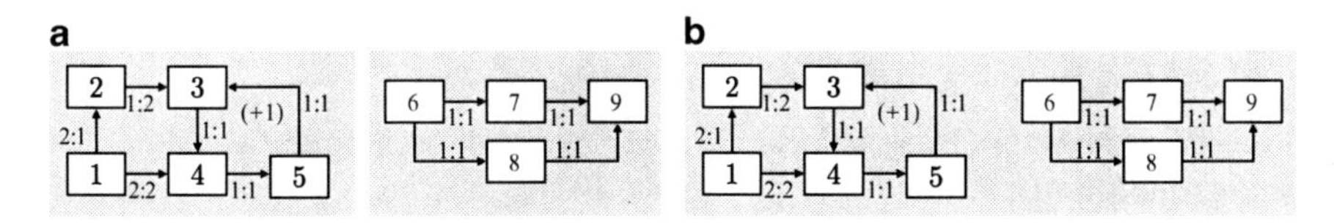

Fig. 6.4 From single to multiapplication scenario. (**a**) Two applications and task graphs. (**b**) Joint representation of two applications

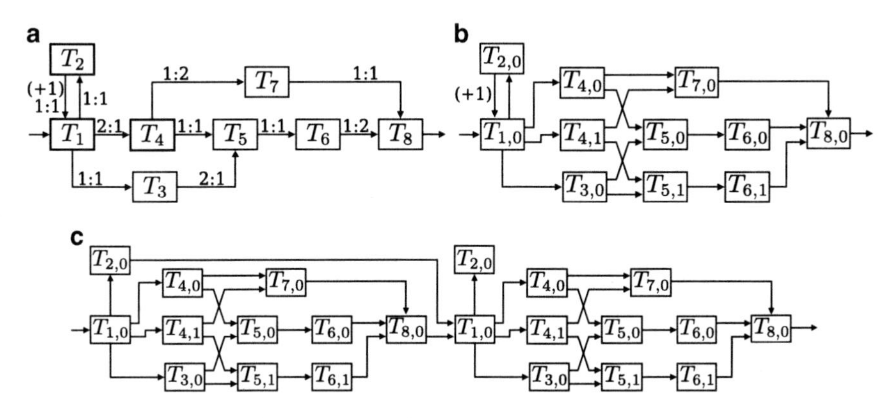

Fig. 6.5 Transformation of the general application task graph to a acyclic directed DFG. (**a**) Initial task graph (TG). (**b**) Respective Feedback Data Flow Graph (FDFG). (**c**) Respective Data Flow Graph (DFG)

1. *Feedback Data Flow Graph (FDFG) generation.* The FDFG is obtained by unrolling the task graph according to the data flow rates so that each data flow is activated once per iteration of the task graph. This condition is true if the rates fulfill $r(d_k) = (1,1), \forall d_k \in D$. Similar graphs are referred to in [20] as homogeneous SDFs. Each original task graph can be transferred to such FDFG by explicit creation of $N(t_i)$ instances of each task. This step leads to an FDFG described as:

$$\text{FDFG} = (T_{\text{FDFG}}, D_{\text{FDFG}}, \delta_{\text{FDFG}}),$$

2. *Directed acyclic Data Flow Graph (DFG) generation.* The FDFG still holds the feedback delay annotations which are essential for later critical path and feedback loop analysis. These must be preserved and transferred to the final graph description. The general definition of the DFG can be given as:

$$\text{DFG} = (T_{\text{DFG}}, D_{\text{DFG}})$$

with T_{DFG} the tasks and D_{DFG} the data flow. The DFG is constructed by unrolling the FDFG several times in parallel

$$\text{DFG} = \{\text{FDFG}_0, \ldots, \text{FDFG}_n \mid \text{FDFG}_i = (T_{\text{FDFG}}, D_{\text{FDFG}} \setminus \{\, d_i : \delta(d_i) > 0\})\},$$

so that $d_k = (t_i, t_j) \in D_{\mathrm{FDFG}}, m = \delta(d_k) > 0$ connect the corresponding instances of t_i in FDFG_l with t_j in FDFG_{l+m} for $l = \{0, ..., n-m\}$.

So far tasks within the graph can be connected with each other, but no in- and output connection is feasible. As applications, like physical layer processing, receive input data, e.g., received data samples from the A/D-converter, and output data to the higher level processing, these in- and output characteristics can be added and are further illustrated with ingoing and outgoing arrows. For practical reasons they are modeled as source and sink tasks so that no loose ends occur within the task graph and no explicit definition of in- and output characteristics is required.

HW Architecture ($x_{\mathbf{Arch}}$)

Current development of complex MPSoC hardware architectures follows the fundamental principle of IP component and platform-based design as already highlighted in Sect. 2.2. Naturally, this design approach produces hardware platforms in the form of block diagrams [258]. In such a block diagram, each vertex/node represents one IP component while directed edges illustrate possible connections between the various components. According to Sect. 2.2 these components can be roughly grouped into the three classes of processing elements (PE), communication architectures (CA), and memories (Mem). A formal specification of such MPSoC hardware architectures can be given as follows.

Definition 6.5 (Hardware Architecture). The MPSoC hardware architecture is defined as a joint system of processing elements, communication architectures, and memories defined similar to [251, 258]:

$$\begin{aligned} HW = (PE, CA, Mem)&: \\ &PE \text{ is a set of processing elements } \{pe_0, ..., pe_p\}, \\ &CA \text{ is a set of communication architectures } \{ca_0, ..., ca_c\}, \\ &Mem \text{ is a set of memories } \{mem_0, ..., mem_M\}. \end{aligned}$$

The interconnection of the hardware platform is defined by the following sets.

- Each processing element is attached to a set of communication architectures [4]

  $$CA(pe_p) = \{ca_i, ..., ca_j\},$$

  or reversible $PE(ca_c)$ defines the connected processing elements onto a communication architecture ca_c

  $$PE(ca_c) = \{pe_i, ..., pe_j\}.$$

[4] Complex NoCs can be either captured as a single communication architecture or can be split into their underlying structure based on network interfaces, communication links, and bridges.

- To conclude the description of the hardware architecture, each memory element is connected to a communication architecture

$$Mem(ca_c) : CA \rightarrow Mem.$$

For later performance evaluation, in addition to the components and their interconnection, other performance characteristics and properties are required. As especially at early design stages only a limited knowledge of the underlying hardware platform may exist, these properties are kept simple during the definition for the mathematical analysis. Detailed aspects and properties can be incorporated or inspected in later simulation when suitable implementation candidates have been identified by the analysis. For a first and crude definition the following basic properties have been identified.

- *Processing Element Properties.* To scale the processing cycles to the time basis, the clock period is required.
- *Communication Architecture Properties.* Similar to the processing element, the clock period scales each clock cycle to the time base. In addition, the data-width defines the number of data elements that can be transferred in a single transaction. Furthermore, the write and read cycles are the number of clock cycles required to write and read a data element.
- *Memory Element Properties.* For practical reasons the latency due to memory accesses is captured within the properties of the communication architecture rather than within the memory element.

With both application and hardware description at hand, the temporal and spatial task mapping is defined.

Temporal and Spatial Task Mapping (x_{Map})

The proposed approach follows the Y-chart principle [223] and the paradigm of orthogonalization of concerns [156]. In this context spatial mapping denotes the distribution of tasks to the different processing elements. Temporal mapping defines the execution with respect to time on a single processing element.

The spatial mapping is restricted by the existence of an interconnect structure between two processing elements formulated as follows.

Definition 6.6 (Restriction of the Spatial Task Mapping[5]). A spatial mapping of a task t_x to a processing element pe_i is given by $PE(t_x) = pe_i$. Therefore, the following condition must be valid for a feasible spatial mapping.

$$\exists ce_c \in CA \ \forall (t_i, t_j) \in D_{\mathrm{DFG}} : pe_i = PE(t_i), pe_j = PE(t_j), pe_i, pe_j \in PE(ce_c)$$

[5] Please note that whenever a processing element is used as a bridge between two communication architectures, it is necessary to insert a bridging task.

Based on this restriction, a valid spatial task mapping is defined as follows.

Definition 6.7 (Spatial Task Mapping).

$STM = (T_{\text{DFG}}, PE(HW))$:
- DFG is the data flow graph of the application $DFG = (T_{\text{DFG}}, D_{\text{DFG}})$
- T_{DFG} are the tasks of the DFG
- HW is the hardware $HW = (PE, CA, Mem)$
- $PE(HW)$ are the processing elements of the hardware
- $\exists PE(t_i) = pe_x, \; \forall t_i \in T_{\text{DFG}}$ with $pe_x \in PE(HW)$
- $\forall (t_i, t_j) \in D_{\text{DFG}}$ definition 6.6 holds.

Figure 6.6 depicts two examples to illustrate valid and invalid spatial task mappings. While the first example violates the last condition with the mapping of task 2 and 3, the second example defines a valid mapping.

The illustrated examples contain structures where multiple tasks are mapped to a single processing element. In such cases temporal resource allocation, also known as scheduling [259], is indispensable to define the execution order of tasks on a shared resource. The following discussion will adhere to the term of *scheduling* to describe the temporal task mapping.

Traditional classification separates *dynamic* and *static* scheduling. Static scheduling defines an exact order of task execution that is determined prior to run-time. Mostly, static schedulers operate on a fixed sequence of tasks, that is repeated continuously until the execution of the application ends. A special case of static scheduling is *Time Division Multiple Access (TDMA)* [260], that has fixed time slices for task execution. When the corresponding time slice finishes, the task is preempted and the next task in sequence continues execution. TDMA-based schedulers find huge importance when latency and throughput constraints need to be guaranteed [261].

In general to compute the optimal or a suboptimal static schedule, the task execution times need to be known and unforeseen dynamic effects should not occur. Besides these issues, computation of a static schedule is typically an NP-complete problem and therefore heuristics are utilized to solve the problem [259].

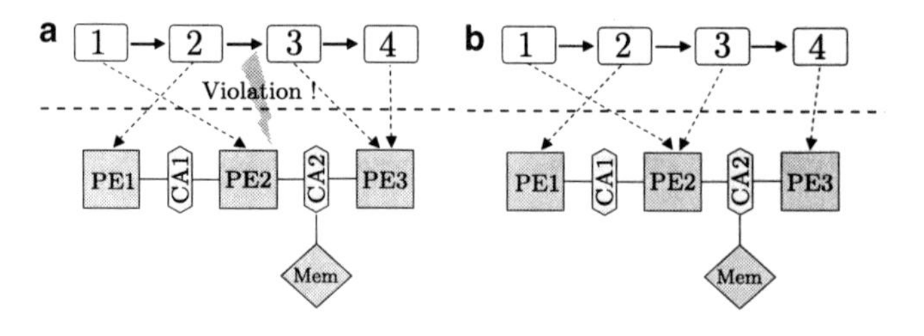

Fig. 6.6 Examples of valid and invalid spatial mappings. (**a**) Invalid Spatial Mapping. (**b**) Valid Spatial Mapping

Because of the restrictions of static scheduling, the technique of dynamic scheduling is beneficial to handle dynamic and unforeseen effects, e.g., to start and stop applications at run-time. It allows run-time optimization of task execution based on a given objective, e.g., priority or load balancing. Unfortunately, the overhead is significant and due to the dynamic task execution, the resulting execution characteristic is commonly nondeterministic. Especially, this makes it difficult or even impossible to use dynamic scheduling in the domain of physical layer processing. In addition, a common characteristic of physical layer applications is the well-ordered structure, in terms of task graphs, that makes static schedules an excellent choice.

The determinism of static scheduling makes analysis in a well-defined mathematical manner possible. In contrast, most approaches based on a mathematical analysis lack support for dynamic scheduling, as nondeterminism and dynamic effects can barely be captured in such approaches. Therefore, investigation of dynamic effects mostly relies on worst-case analysis or simulation-based approaches, like the one later discussed in Chap. 7.

Considering static scheduling, the temporal resource allocation, i.e., an ordered list of task executions on a processing element, is defined as follows.

Definition 6.8 (Static Scheduling). A static schedule defines the order of task execution on a processing element $SC(pe) = (t_0, ..., t_n)$. A condensed representation of a static schedule can be formalized as a *Control Flow Graph (CFG)*.

$$
\begin{aligned}
CFG = (T_{CFG}, C_{CFG})&: \\
&T_{CFG} \text{ is the set of tasks } \{t_1, ..., t_n\}, \\
&C_{CFG} \text{ is a set of control dependency edges } \{(t_i, t_j) : t_i, t_j \in T_{CFG}\}, \\
&\forall (t_i, t_j) \in C_{CFG} : \text{ there is no path } (t_j, ..., t_i), \\
&\forall t_i \in T_{CFG} : |\{(t_i, t_j) \in C_{CFG}\}| \leq 1, |\{(t_j, t_i) \in C_{CFG}\}| \leq 1
\end{aligned}
$$

The list order implies task dependencies of $t_i \in T_{CFG}$ on the finish of the proceeding task executions $t_0, ..., t_{i-1} \in T_{CFG}$.

To support investigations of different schedulers, the proposed analysis framework includes a condensed language to specify arbitrary static schedules. This grammar is briefly highlighted in Appendix A.2.

To join the application's DFG (Definition 6.4) and the CFG of the scheduling (Definition 6.8) with respect to the hardware architecture (Definition 6.5), the set of tasks $\{T_{CFG}(pe) \forall pe \in PE(HW)\}$ and $\{T_{\mathrm{DFG}}\}$ must be equivalent. Hence, if the DFG is constructed by unrolling the original task graph, the CFG must be equally converted by task duplication and unrolling.

Additionally, the spatial mapping denotes the processing element that executes the task. This inherently defines the task execution and communication characteristic, which can be captured at early design stages within a random variable $X_i = X(t_i, pe_j)$ allowing further analysis. The stochastic task execution characteristic is discussed in depth later.

Stochastic Description ($X_i \sim (t_i, pe_j)$)

The spatial task mapping (STM) determines the underlying processing element $pe_j = PE(t_i)$ that is utilized to execute a given task t_i. Together the task implementation, e.g., the software code, and the underlying processing element pe_j determine the execution characteristic $X_i = X(t_i, pe_j)$ of the implemented task.

When considering the final implementation, this task execution characteristic X_i might be given as a single value that equals the execution time. Since the proposed analysis targets early design phases, usually the final implementation is not available and therefore cannot be determined that precisely. Instead, system architects need to perform the analysis based on characteristics that tend to incorporate uncertainties. These uncertainties are generally due to unknown implementation knowledge or imprecise knowledge of the system behavior (Fig. 6.7).

Implementation Knowledge. While particular implementations are already present, others are only known from an algorithmic perspective with a rough idea of the later implementation. In the first case a precise execution characteristic can be computed or measured, while in the second case only estimates can be applied.

System Behavior. With imprecise knowledge of the system environment, performance characteristics of a task cannot be given without an uncertainty. For example, capturing a cache behavior or bus contention precisely is highly complex, while these can be captured stochastically with sufficient precision.

To deal with the occurrence of such uncertainties and imprecise implementation knowledge at early design phases, the task execution characteristic is stochastically defined as a *random variable*.

Definition 6.9 (Stochastic Task Execution Characteristic). The task execution characteristic $X_i(t_i, pe_j)$ is determined by a *random variable* that follows a *probability density function (pdf)* $f_{X_i}(t)$, a *cumulative distribution function (cdf)* $F_{X_i}(t)$ or other statistical function descriptions. As the pdf and cdf are the most common descriptions, only these will be used in the following. Based on the description of the task execution characteristic with the pdf, the random variable $X_i(t_i, pe_j)$ is given by $f : \mathbb{R} \to \mathbb{R}_{>0}$, $x \mapsto f_{X_i}(t)$.

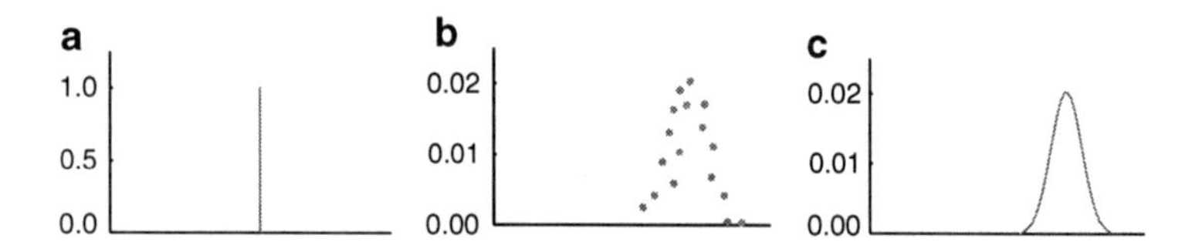

Fig. 6.7 Different types of stochastic parameter description illustrated by their probability density functions. (**a**) Perfect Knowledge. (**b**) Simulation Results. (**c**) Stochastic Description

To minimize the error with respect to the final implementation, the random variable should at best match the exact execution characteristic. However, with limited knowledge and no final implementation at hand, system architects have to estimate the task execution. Other methods to evaluate these characteristics can be based on profiling or measurement results. Figure 6.7 exemplifies these different possible scenarios. The first case illustrates perfect knowledge of the implementation that relates to a fixed and given implementation. Interpolation of simulation results can lead to another pdf as depicted in the second example, while the third one completely relies on expert knowledge. Designers should compensate their unreliable knowledge by choosing more wide spread distributions, whereas with detailed knowledge more narrow distributions may be selected. However, this should be treated with the highest degree of caution, since wrong estimates might lead to false design decisions.

The ultimate goal is to identify the exact execution characteristics that reflect the final implementation, which - in turn - means the implementation of the complete system. To achieve this an iterative refinement process is defined as sketched in the overview (Fig. 5.1).

Within the early exploration, the mathematical analysis is performed subject to the given constraints.

Constraints ($\underline{c}$)

To evaluate whether a system operates properly or not, the hard constraints of latency and throughput must be preserved for applications, especially for physical layer ones. These constraints mostly occur in so-called critical paths and feedback loops. In the following discussion only critical paths are going to be referred to, as the feedback loops define special cases of critical paths.

Definition 6.10 (Critical Path). A critical path (CP) is a sequence of tasks within the task graph $TG = (T_{TG}, D_{TG})$, the FDFG, or the DFG of the application. The definition based on the original task graph is as follows

$$CP = (t_s, ..., t_w, ..., t_e) : t_s, ..., t_w, ..., t_e \in T_{TG}$$

Furthermore, the condition must hold that for a given critical path each task on the way within the sequence from the start point t_s to the end point t_e must be connected. Please note when considering the FDFG and DFG not all tasks in the t_s, t_e-path must be part of the critical path itself.

Failure to meet the given constraints most likely leads to nonoperational devices that fail compliance tests. Hence, evaluation of these constraints is a key proposition of the analysis. The subsequent section will introduce the analysis algorithm that forms the centerpiece of the mathematical implementation model to analyze whether the constraints are met or not.

6.2 Analysis Algorithm

The heart of the proposed analysis is an algorithm that operates on graphs. It is based on general graph theory and utilizes the stochastic task description to evaluate the performance characteristics of critical paths subject to evaluation. A vehicle utilized for analysis is the so-called *Analysis Graph (AG)*. This analysis graph preserves the dependency structure of the application and temporal and spatial task mapping, which inherently contains the underlying hardware architecture.

The analysis can be structured into four sequential steps also depicted in Algorithm 1.

1. *Analysis Graph Calculation.* Both data- and control-flow form the foundation of the analysis graph. While the DFG of the applications (Definition 6.4) holds the data-flow, the control-flow is preserved in the schedules of the processing elements (Definition 6.8). Subsequently, merging both graphs generates the analysis graph.
2. *Analysis Graph Simplification.* To keep complexity reasonable, the analysis graph is simplified by means of pure standard graph reduction as found in [257]. All necessary information is preserved to later compute the critical path performance. This comprises data- and control- dependencies.
3. *Analysis Precalculation.* Prior to the final evaluation of the critical paths, precalculations need to be performed.
4. *Critical Path Evaluation.* Finally, the critical paths are evaluated based on the stochastic task execution characteristics which depends on the individual task and its spatial mapping.

By merging the control- and data-flow graphs, the analysis graph holds all dependencies. The characteristic of a critical path is computed by simply adding

Algorithm 1 Functional Overview of the algorithm

Input: Application Task Graph TG, Hardware HW, Schedules SC, Critical Paths CP
Output: Critical Path Characteristics $\{X_{CP_i}\}$

// Analysis Graph Calculation (Sect. 6.2.1)
1 N = DetermineActivation(TG);
2 ($FDFG, SC$) = DuplicateTasks(TG, SC);
3 (DFG, CFG) = UnrollIterations($FDFG, SC$);
4 AG = ConstructAnalysisGraph(CFG, DFG);

// Analysis Graph Simplification (see Appendix A.1)
5 AG = MergeVertices(AG);
6 AG = EliminateShortcuts(AG);
7 AG = MergeVertices(AG);

// Analysis Precalculation (Sect. 6.2.2)
8 AG = CalculateDependencyDelays(AG);

// Critical Path Evaluation (Sect. 6.2.3)
9 $\{X_{CP_i}\}$ = EvaluateCriticalPaths(AG, CP);

the random variables of all nodes within the path. Unfortunately, dependencies due to control- and data-flow *must* be considered, since they typically prohibit such simple calculation. These exist due to the nature of parallel processing on multiple resources and will later be discussed in-depth.

So far the overall steps of the analysis algorithm that are subsequently discussed have been sketched.

6.2.1 Analysis Graph Calculation

A joint and comprehensive representation of control- and data-flow dependencies in form of an analysis graph (*AG*) allows efficient computation of the critical path characteristics. The analysis graph is constructed so that it comprises all relevant information as well as the data- and control-flow dependencies. As discussed earlier in this chapter, the task graph (*DFG*) holds the data-flow characteristics, while the temporal and spatial task mapping comprises the control-flow information stored comprehensively in a control flow graph (*CFG*).[6] Since these graphs are either derived from the initial application task graph (*TG*) (Definition 6.4) or the static schedules ($SC(pe_a)$) (Definition 6.8) these first need to be constructed. Joining both, *DFG* and *CFG*, into the unique analysis graph serves as the basis for the critical path evaluation. Please note that developers usually start with the initial task graph. However, more fine-grained inspection of scheduling techniques might require direct modifications of the *CFG* to achieve superior performance by optimizing the schedule. In general the computation of the *AG* consists of the subsequent steps.

1. Reduce data-flow rates to a 1-to-1 relation by duplicating tasks and edges ($TG \rightarrow FDFG, SC \rightarrow SC$).
2. Remove data-flow delay annotations by instantiating multiple iterations ($FDFG \rightarrow DFG, SC \rightarrow CFG$).
3. Construct *AG* by merging data- and control-flow graphs ($DFG \,\&\, CFG \rightarrow AG$).

In Sect. 6.1.2 the transformation of the initial application task graph (*TG*) to the directed acyclic data-flow graph (*DFG*) has already been introduced. Figure 6.5 exemplifies this for the given task graph TG_0.

As a first step in this generation, the task activation rate $N(t_i)$ of each task is computed with respect to the in- and output data rates. The utilized algorithm for computing these activation rates is based on the solver algorithm introduced in [87].

With the obtained activation rates, the tasks and edges are duplicated to construct the FDFG. When duplicating the tasks $N(t_i)$-times, the edge that connects two tasks t_i and t_j with the rate (r_i, r_j), the edge needs to be instantiated $R \cdot lcm(r_i, r_j)$, $R \in \mathbb{N}^+$ times, with *lcm* being the least common multiple.

[6] Control flow, e.g., the control flow from the MAC to PHY layer, can be captured either in the DFG as data-flow representation or in the CFG as a schedule.

The final step in retrieving the DFG is to unroll the FDFG to eliminate existing delay dependencies, which typically occur in feedback loops, i.e., due to the delay length of a propagated data element. The number of times the FDFG needs to be unrolled is one plus the maximum delay dependency within the FDFG and the initial TG, that is

$$N_{iter} = 1 + max(\{\delta_{\text{FDFG}}(t_i) : \forall t_i \in T_{\text{FDFG}}\})$$

To form the analysis graph (AG), the retrieved DFG needs to be joined with the CFG that can be computed based on the set of schedules of each processing element $SC(pe_a)$ within the hardware architecture (HW).

The initial schedules $SC(pe_a)$ for each processing element need to be transformed to reflect the multiple task instances and iterations. Hence, on the basis of the determined activation rates, the multiple instances of the task need to be added to the scheduling. For example, a schedule (T_3, T_4, T_6, T_8) can be transferred to $(T_{3,0}, T_{4,0}, T_{4,1}, T_{6,0}, T_{6_1}, T_{8,0})$, $(T_{3,0}, T_{4,0}, T_{6,0}, T_{4,1}, T_{6_0}, T_{8,0})$, or any other schedule that respects the data-flow dependencies. Within the proposed framework, the default scheduling policy is based on *self-time schedule* policy [262]. Since the selected scheduling can have a significant impact on the overall system performance, more fine-grained inspection is supported by the analysis workbench as further discussed in Appendix A.2.

Iterations due to delay dependencies request repetition of the schedules so that the schedules execute N_{iter}-times. Finally, all schedules are combined in a CFG that contains M disconnected subgraphs with M equal to the number of processing elements being part of the hardware architecture.

Finally, the AG can be computed based on the DFG and CFG. Given the CFG, the data-flow edges of the DFG are added. Connecting two disconnected graphs of the CFG by a data-flow edge defines a data transfer from one processing element to another one. Hence, the analysis needs to identify directly if a communication architecture that can handle the transfer exists. If no suitable communication architecture is available, the spatial mapping is invalid as it violates the mapping restriction given by Definition 6.6. In case a communication architecture exists, vertices need to be added that reflect the data *write* and *read* transaction. The corresponding data-transfer time depends on the size of the data and the underlying communication architecture.

To exemplify the Analysis Graph generation, the scenario depicted in Fig. 6.8 is investigated. The initial task graph can be transformed to a DFG as previously shown in Fig. 6.5. The temporal and spatial task mapping is defined by the two schedules of the processing elements pe_a and pe_b.

$$SC(pe_a) = (T_1, T_2, T_5, T_7)$$
$$SC(pe_b) = (T_3, T_4, T_6, T_8)$$

Figure 6.9 highlights the analysis graph generation as discussed in-depth. First, the CFG and DFG are joined (Fig. 6.9a) and second, the vertices are added where communication external to the current processing element occurs (Fig. 6.9b). As the

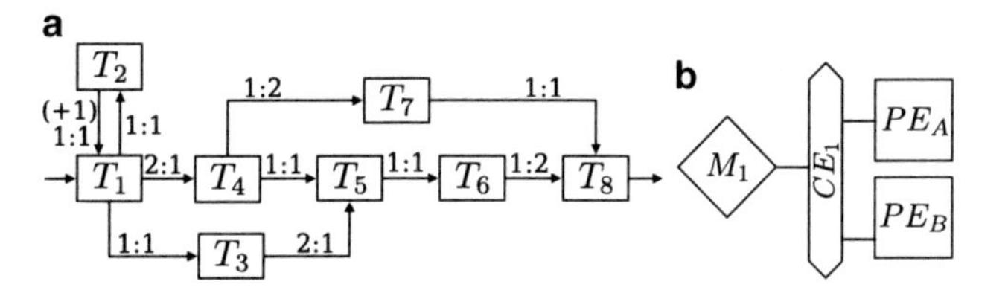

Fig. 6.8 Example: application and hardware architecture. (**a**) Initial task graph (TG). (**b**) HW architecture

processing elements perform the data transfer and meanwhile they are blocked, the communication vertices need to be included into the processing element schedules, respectively, the CFG part (Fig. 6.9c). To simplify later analysis, graph reduction, especially edge reduction, is necessary to keep the computational complexity low (Fig. 6.9d).

So far an analysis graph has been constructed on which the analysis, in particular the evaluation of the critical paths, is performed. Before introducing the final critical path evaluation, a precalculation to simplify the evaluation is introduced.

6.2.2 Analysis Precalculation

The evaluation of the critical paths (*CP*) is based on the analysis graph (*AG*) and the stochastic description of each task. A more generalized definition of the critical paths than the one in Definition 6.10 is a path from the starting vertex to an end

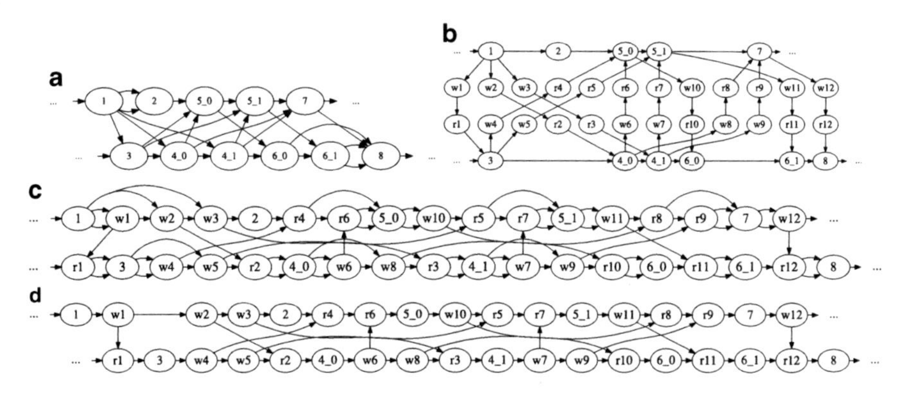

Fig. 6.9 Evaluation of the analysis graph. (**a**) The joint DFG and CFG. (**b**) Adding read and write communication vertices and edge reduction. (**c**) Insert communication into schedules. (**d**) Edge reduction

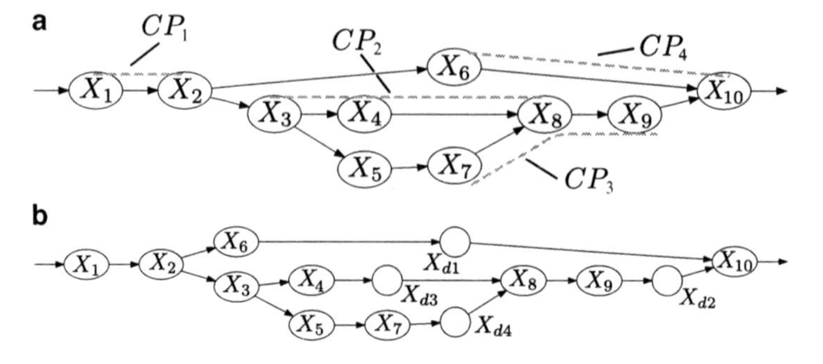

Fig. 6.10 Example: analysis graph with exemplary critical paths and dependency delays. (**a**) Critical paths in an analysis graph. (**b**) Inserted dependency delays

vertex with arbitrary vertices on the way. Hence, the critical path consists of vertices each describing a task which, in turn, is characterized by a random variable $X_i(t_i, pe_j)$. According to Definition 6.9 the random variable depends on the task and the processing element which executes the task according to the spatial task mapping (Definition 6.7). The main objective is to calculate the execution characteristic of the critical path X_{CP_i}.

The simplest possible critical path is one that consists of one task $CP_i = (T_i)$. The execution characteristic equals the task characteristic, i.e., $X_{CP_i} = X_{T_i}$. In general, evaluation is not that simple and requires much more effort as discussed in the following. Figure 6.10 visualizes the main issues and differences that occur while computing the critical path characteristics.

Figure 6.10a depicts four exemplary critical paths. While the CP_1 and CP_2 are rather simple to evaluate, CP_3 and CP_4 require much more effort, induced by control- and data-flow dependencies. The sequential execution of task t_1 and t_2 defines the critical path CP_1. Apparently the random variable X_{CP_1} is the addition of the random variables X_1 and X_2. With the random variables X_1 and X_2 statistically independent, the probability density function of X_{CP_1} is the convolution of the probability density functions of the random variables X_1 and X_2 (c.f. [263, page 193f.]).

In contrast to the simple addition for CP_1 the critical path $CP_2 = (T_3, T_4, T_8)$ is influenced by the parallel path (T_3, T_5, T_7, T_8) since the dependency edge (T_7, T_8) prevents execution prior to the finish of the parallel path execution. Hence, the random variable of the critical path CP_2 computes to $X_{CP_2} = max(X_{CP'_2}, X_{CP''_2})$. The probability density function of the CPs random variable computes to

$$f_{X_{CP_2}}(t) = f_{X_{CP'_2}}(t) F_{X_{CP''_2}}(t) + F_{X_{CP'_2}}(t) f_{X_{CP''_2}}(t)$$

when statistical independence of $X_{CP'_2}$ and $X_{CP''_2}$ is given. With respect to the given example, this only holds if the start task (T_3) and the end task (T_8) are not considered in the maximum operation. Since these tasks are common for both paths, the random variables can simply be added to the result of the maximum operation so that independence for the maximum operation is given.

So far the introduced techniques to compute the critical path characteristics are not sufficient to analyze the critical paths CP_3 and CP_4. Following the general approach of computing the random variable X_{CP_3} by adding the random variables along the path, would lead to $X_{CP_3} = X_7 + X_8 + X_9$. In fact this result is *wrong* as a delay occurs between execution of T_7 and T_8 when the sum of execution times of X_5 and X_7 is larger than the one of X_4. Stochastically this delay is determined by:

$$\begin{aligned} X_{\text{sub}} &= X_4 - (X_5 + X_7) \\ X_{d4} &= max(0, X_{\text{sub}}) \end{aligned}$$

In contrast, execution of T_8 might be delayed when the realization of T_5 and T_7 is larger than the one of T_4. Hence, the delay between the nodes of X_4 and X_8 is captured by X_{d3} as:

$$\begin{aligned} X_{\text{sub}} &= (X_5 + X_7) - X_4 \\ X_{d3} &= max(0, X_{\text{sub}}) \end{aligned}$$

A negative delay of the random variable X_{sub} defines that the other path is delayed and no delay at the investigated path occurs. Hence, the delay is zero in such cases. This is reflected by a probability density function that has a dirac delta function at zero with the weight of $\int_{-\inf}^{0} p_{X_{\text{sub}}}(t)\mathrm{d}t$ and the positive part of the distribution of the random variable retrieved by substraction. The retrieved random variable of this calculation is a so-called *dependency delay*. Dependency delays need to be considered and applied whenever a stall may appear, which in general occurs at vertices that are *joins*. A *join* is a vertex that has multiple incoming edges and the set of joins V_J is

$$V_J = \{v \in V(AG) : |\{e = (v_t, v_h) \in E(AG) : v_t, v_h \in V(AG), v_h = v\}| > 1\}$$

Since dependency delays might occur in all incoming edges of the vertices $v_x \in V_J$, for each join $v_x \in V_J$ with incoming edges $e_0, ..., e_N$ a dependency delay X_{di} needs to be computed for each edge $e_i, i = 0, ..., N$. To compute them, all random variables $X_{p_0}, ..., X_{p_n}$ that describe the path to the incoming edge e_i need to be considered. Then the dependency delay X_{di} can be determined by

$$\begin{aligned} X_{max} &= max(X_{p_0}, ..., X_{p_{(i-1)}}, X_{p_{(i+1)}}, ..., X_{p_n}) \\ X_{\text{sub}} &= X_{max} - X_{p_i} \\ X_{di} &= max(0, X_{\text{sub}}) \end{aligned}$$

For this operation all paths $p_0, ..., p_n$ must have the same starting vertex. Such starting point must be a vertex with multiple outgoing edges, that is a so-called *split*. Similar to the set of joins V_J the set of splits V_S is defined as

$$V_S = \{v \in V(AG) : |\{e = (v_t, v_h) \in E(AG) : v_t, v_h \in V(AG), v_t = v\}| > 1\}$$

To compute the execution characteristics of the paths $p_0, ..., p_n$, the common *split* vertex $v_s \in V_S$ needs to be identified by a backward search starting at the *join* vertex $v_j \in V_J$. Exactly this algorithm was applied in the previous example of X_{d_3}.

For the example in Fig. 6.10a, the additional data-dependency delays X_{d_1}, X_{d_2}, and X_{d_4} need to be calculated. Their detailed computation and the algorithm used are discussed in detail within Appendix A.3. Finally, with the dependency delays included into to the analysis graph (*AG*), all critical paths can be evaluated as discussed in the following.

6.2.3 Critical Path Evaluation

In general, each critical path can now be simply determined by adding the random variables along the path. For example, the random variable of critical path CP_2 depicted in Fig. 6.10a computes to $X_{CP_2} = X_7 + X_{d4} + X_8 + X_9$. As previously stated, the addition of random variables means a convolution of their probability density functions when statistical independence is ensured. For the given scenario, the random variable $X_{d4} = max(0, X_4 - (X_5 + X_7))$ is no longer independent due to the random variable X_7. This violates the assumption of independence between X_{d4} and X_{CP_2} since they are dependent. However, the induced error is minor compared to the uncertainties due to imprecise implementation knowledge at early design stages. A brief evaluation of the encountered error is given in Appendix A.4. Due to the occurring errors in such cases, critical paths are computed without the use of dependency delays whenever feasible, like for the exemplary critical paths CP_1 and CP_2.

It should be noted that the current implementation of the analysis cannot handle critical path calculations per bit when bit reordering appears, e.g., interleaver and deinterleaver operations. Instead analysis of critical paths can be performed blockwise.

In the overview (Chap. 5), the result interpretation of critical paths has already been discussed. Furthermore, during the case study, detailed analysis results will be given and discussed in depth.

6.3 Simulation Link and Back Annotation

Both analytical and simulation models have their own advantages and disadvantages. Typically these approaches are completely separated, which requires twice the development effort in two different environments. To bridge this gap in MPSoC design flows, the proposed analytical analysis is integrated into a simulation-based framework. Figure 6.11 depicts the refinement flow from the mathematical implementation model down to the abstract simulation model.

Accordingly, at the start of the design process an analytical implementation model is developed based on the previously discussed analytical workbench.

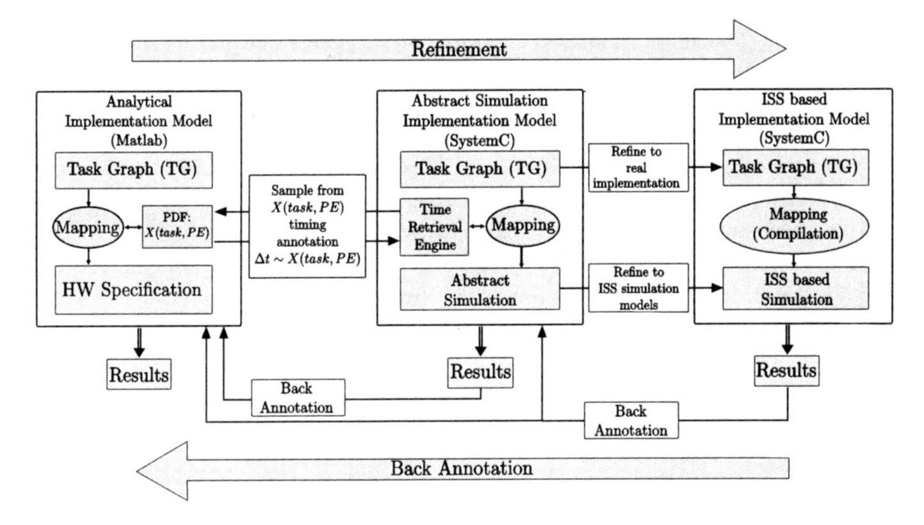

Fig. 6.11 Mathematical to ISS refinement of the implementation model

This computes the characteristic X_{CPi} for each critical path, based on the individual task characteristics $X(task, PE)$. If a suitable candidate has been identified, developers can refine the analytical model to a SystemC simulation model based on the framework discussed in [250]. The central element in this framework is an *abstract processor simulator* called VPU operating on the *annotation of execution characteristics* paradigm to reflect the task behavior. This smooth transition is enabled by the *Time Retrieval Engine* which samples from the pdf $X(task, PE)$ a timing annotation Δt at run-time ($\Delta t \sim X(task, PE)$). This retrieved value is then propagated to the VPU which annotates this. Thus, *without any* modification, the analytical implementation model can be evaluated in a simulation-based framework. Encountered effects within the simulation can be iteratively back annotated to the analytical implementation model.

The demand for such back annotation has its origin in two main sources of inaccuracies. On the one hand, dynamic effects that are caused by conflicting memory accesses make it hard to predict the exact timing behavior of communication requests. On the other hand, inaccuracies will be caused by false estimates of the execution times due to limited implementation knowledge in early design steps.

The principle of back annotation is rather simple and allows for a quick update of the analytical model. Whenever a mismatch between the analytical and a lower level implementation model is recognized, the more precise execution characteristic is measured at a lower abstraction level. This can either be a task execution or a data transfer on a shared communication architecture. The measured execution characteristic is summarized as a random variable with a particular pdf $X(task, PE)$ or $X(data, CA)$. Finally, the earlier utilized random variable is updated, hence, the inaccuracy is removed and system architects can resume the top-down design approach.

Since the final implementation is naturally the goal of such a design process, developers can refine the abstract simulation model further to an implementation model based on an Instruction Set Simulator (ISS). This step comprises implementation of the software, mostly in C programming language and replacement of the abstract processor simulator by an ISS. Such simulation models allow measurement of the actual performance in terms of instruction- or cycle-accurate behavior. Again the obtained results can be back annotated to the abstract simulation and mathematical implementation model. Here, as previously discussed, the uncertainty due to the unknown is reduced by utilizing the actual measured performance. This enhances the estimates and provides results, which are closer to the real implementation in later design iterations.

Chapter 7
Abstract Simulation Implementation Model

Leaving the domain of analytical-based design space exploration, the discussion now turns to the utilized abstract simulation. First an overview of the underlying technique along with its main components is given. The key element, the Virtual Processing Unit (VPU) technology, is presented in-depth. Finally, this chapter concludes with the introduction of the refinement flow from abstract simulation down to an implementation model based on Instruction Set Simulators (ISSs).

7.1 Overview and Key Components

The previous analysis of applications has highlighted the diverse structures and demands of applications for wireless communication devices. Within this discussion, two main application classes have been identified. The first class covers the domain of physical layer processing and multimedia applications. Thanks to their regular structure these can be partitioned into task graphs and can be efficiently modeled as process networks, e.g., Kahn Process Networks (KPN) and Synchronous Data Flows (SDFs). This eases the use of programming models which can provide graphical design entries for software visualization (Chap. 6). Examples of such representations are SDF graphs, UML 2.0 activity diagrams, component-based software design, and other programming models based on graph structures to be discussed later in Sect. 7.5.2. In addition, computation of static schedules at compile time ensures a deterministic behavior and minimizes the occurring overhead as no dynamic scheduling is encountered, e.g., based on an operating system.

In contrast, the second class of applications makes heavy use of dynamic scheduling techniques since these need to respond to nondeterministic user interactions. Especially, in this domain operating systems or real-time operating systems are frequently utilized besides other software abstraction levels such as middlewares for abstract data communication like MCAPI [264]. User interactions and effects of dynamic scheduling prohibit a simple software visualization. One approach to consistently describe such applications in a graphical fashion are the interaction overview diagrams of UML 2.0 [27]. However, the occurring overhead in terms of performance loss can be significant when using these modeling approaches.

T. Kempf et al., *Multiprocessor Systems on Chip: Design Space Exploration*,
DOI 10.1007/978-1-4419-8153-0_7,

For development of a complete MPSoC platform, these various applications need to be jointly implemented on a single device. This results in the need for a combined design methodology to satisfy all requirements. As a consequence these different requirements imply many challenges for the abstract simulation model.

Such a simulation model needs to support static scheduling and arbitrary dynamic scheduling techniques. Especially in the presence of dynamic scheduling, simulation-based approaches are superior to analytical models, which are either not applicable or have severe issues in modeling these scenarios. One of such simulation-based approaches is the framework built upon the previously sketched VPU concept. The following discussion of the abstract simulation model focuses more on the system-level aspect and the practical use, while a detailed operational semantic can be found in [100, 250].

7.2 Virtual Processing Unit Concept

Intention of the Virtual Processing Unit (VPU) is to mimic the behavior of any kind of processing element from general purpose processor core to highly specialized hardware accelerators. This creates serious challenges for the underlying VPU simulation model as it has to cope with a wide range of software and hardware issues.

Figure 7.1 illustrates the four main scenarios induced by the software and hardware. Necessarily, they have been considered during specification and implementation of the VPU. According to their individual characteristic these are classified as follows.

- *Processor core with single-threaded application.* This scenario defines what refers to the traditional processor core use-case. Here a single application exclusively executes on the underlying processor core. Since C programming language dominates software development for embedded devices, it denotes the execution of a simple C *main* function. Typically, applications make use of libraries and device drivers, e.g., to access peripheral devices like a display controller.
- *Processor core with multithreaded application.* Apart from the simple single-threaded use-case, the multithreaded execution of applications obligates the use of an operating system or a small scheduling kernel that allows switching between applications. In addition to the use of operating systems, such scenarios often utilize middlewares and multiprocessor communication layers like the Multicore Association Communication API (MCAPI) [264] and Polycore's Poly-Messenger [265].
- *Programmable hardware accelerator.* Hardware accelerators are used to speed-up the execution of one particular function or a set of functions in general. Usually their behavior is reactive, like a coprocessor. Programmable ones execute rather small and dedicated programs also called firmwares. This programmability enables minor software updates in the field.

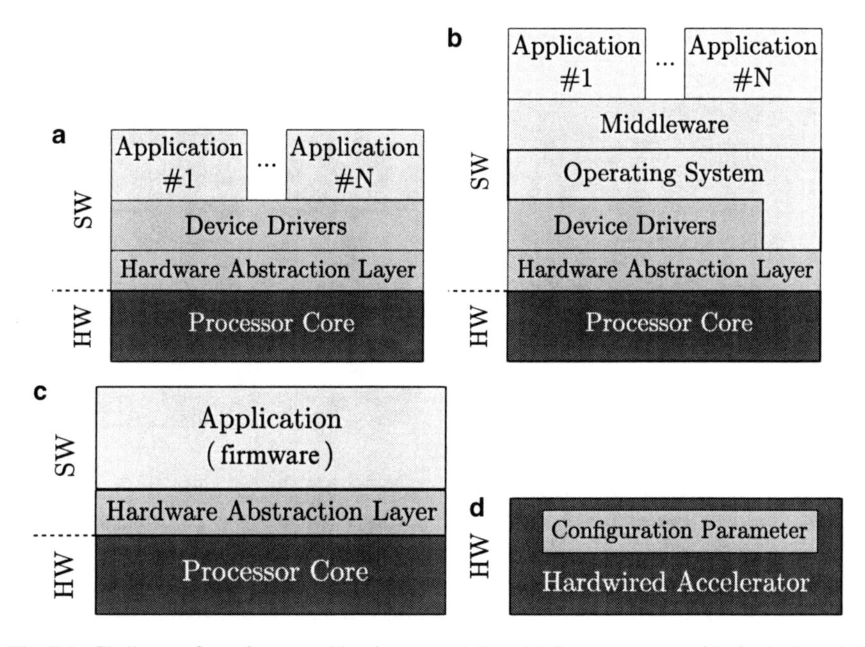

Fig. 7.1 Challenges for software and hardware modeling. (**a**) Processor core with single-threaded application. (**b**) Processor core with multi-threaded application. (**c**) Programmable hardware accelerator. (**d**) Hardware accelerator

- *Hardwired accelerator.* In contrast to programmable hardware accelerators, hardwired ones can only be configured. Therefore, these provide less postfabrication flexibility as only configuration parameters can be set and updates cannot be applied.

Separation between these different classes is not necessarily too strict. For example, in TI OMAP platforms typically the digital signal processor (DSP) behaves similarly to a coprocessor, by executing a function when requested. However, a DSP bios allows support for multithreaded execution of applications on the DSP if required [266].

Besides these software related issues, a huge number of hardware features exists to improve the performance of a single processor core. The basic classification of computer architecture by Flynn in 1966 [46] only considered four different classes, namely Single Instruction Single Data (SISD), Single Instruction Multiple Data (SIMD), Multiple Instruction Single Data (MISD), and Multiple Instruction Multiple Data (MIMD). Enabled by latest silicon and architecture advantages, the advent of new techniques such as Very Long Instruction Words (VLIW), Super-Pipelining, Superscalar, Hyper-threading, etc. has constantly increased performance at the cost of increasing architecture complexity. To define a general and abstract processor simulation model, the VPU model has been envisioned on the principle of *execution characteristic annotation*. This method allows capturing the behavior of

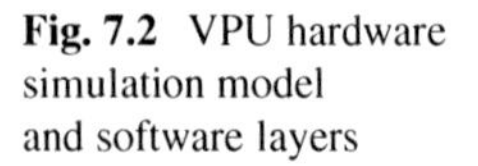

Fig. 7.2 VPU hardware simulation model and software layers

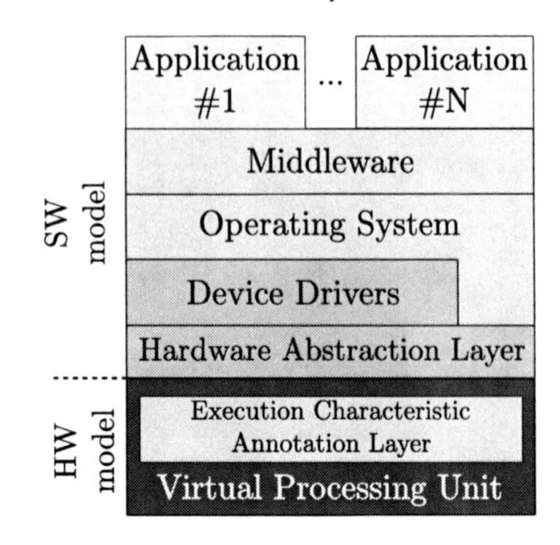

arbitrary hardware effects along with concurrent execution of multiple applications. Recent and past research has shown that accurate modeling is possible. However, modeling of the behavior is quite sensitive to the underlying hardware architecture and its provided features. For example, the later case study discusses this issue for the class of general purpose and digital signal processing processor cores in detail (Sect. 8.1). To support these divers hardware features, *different key concepts* for various annotation techniques have been introduced to cover all coarse- and fine-grained aspects. The next section discusses the fundamentals of the task-based annotation layer.

Apart from the execution characterization, the VPU supports the discussed software features with a layered simulation model as depicted in Fig. 7.2. The underlying hardware is captured by the VPU, which inherently includes aspects like automatic memory address resolution. Inspired by traditional software development, a hardware abstraction layer, device drivers, an operating system, a middleware, and an application layer have been defined (Fig. 7.2). All of these layers operate on top of the VPU simulation model that inherently includes the concept of characterizing the execution behavior especially with respect to the timing.

At the lowest level of detail, the hardware abstraction layer (HAL) of the VPU provides functionality to address external memories and peripherals as later highlighted in Sect. 7.4.1. For example, bus or Network-on-Chip (NoC) centric communication primitives operate on the memory mapped I/O principle [24] that enables simple software development of device drivers and other external VPU communication.

Equal to device drivers of typical processor cores, the device drivers of the VPU can access arbitrary components like peripherals connected to bus or Network-on-Chip (NoC) architectures over external communication ports. Implementation of such device drivers can differ significantly. Accordingly, these include the annotations of execution characteristics to capture the behavior of the driver execution itself. To expose the concept of such drivers Sect. 7.4.2 discusses the fundamental principle, and sketches existing examples such as memory mapped, interrupt-driven, and DMA-driven I/O devices.

The operating system layer is placed above the elementary device drivers. Besides the modeling of different hardware architectures, the generic VPU model has to consider a wide range of operating systems and real-time operating systems. For this reason the operating system is kept highly generic and aims at the common characteristics of all available ones; that is the process management and interprocess communication and synchronization. This includes dynamic task creation and termination along with other operations such as the starting and stopping of tasks. The principle and supported features of the generic OS are documented in Sect. 7.4.3.

Above the operating system layer middlewares are deployed which typically include services like efficient processor-to-processor data communication. The current implementation executing on the VPU provides only the *Task Dispatcher* middleware (Sect. 7.4.4). However, the clear structure and interface definition allows quick implementation of additional layers.

Placed at the top, the application layer supports the execution of one or multiple applications. Using a static scheduling can neglect the middleware and operating system layers, whereas using the operating system enables dynamic-scheduling techniques. In such cases developers can select either to use the graphical design entry of the task-based programming model if applicable or can implement the application in textual mode. Whether choosing one or the other, the task-level annotation remains equal and developers can utilize one of the available techniques discussed next.

7.3 Annotation Principle of Execution Characteristics

At early design stages the knowledge of the algorithms to be implemented and of the underlying hardware differs significantly. For example, algorithms like FIR filters are well known and execution characteristics of an existing implementation can directly be given. In contrast, other algorithms might be completely new and unknown from the implementation level at the start of the design cycle. Accordingly, the underlying architecture is yet unknown and development is still ongoing. Therefore, only coarse-grained estimates of the corresponding execution characteristic are available. Naturally for all exploration techniques, precision of the results relies on the input parameters. Clearly this demands incorporating the most precise available characteristics of both application and architecture at each design stage. To keep track of the large range of implementation knowledge, the abstract VPU simulation model facilitates various annotation techniques. These range from fine- to coarse-grained annotations of the execution behavior.

Figure 7.3 depicts the covered annotation techniques and implementations relating to the implementation knowledge. While the x-axis captures the applicable annotation methods, the y-axis of the chart illustrates the available functional implementation options. Starting with pure traffic generators containing no functional implementation, the range is determined in multiple steps till the final

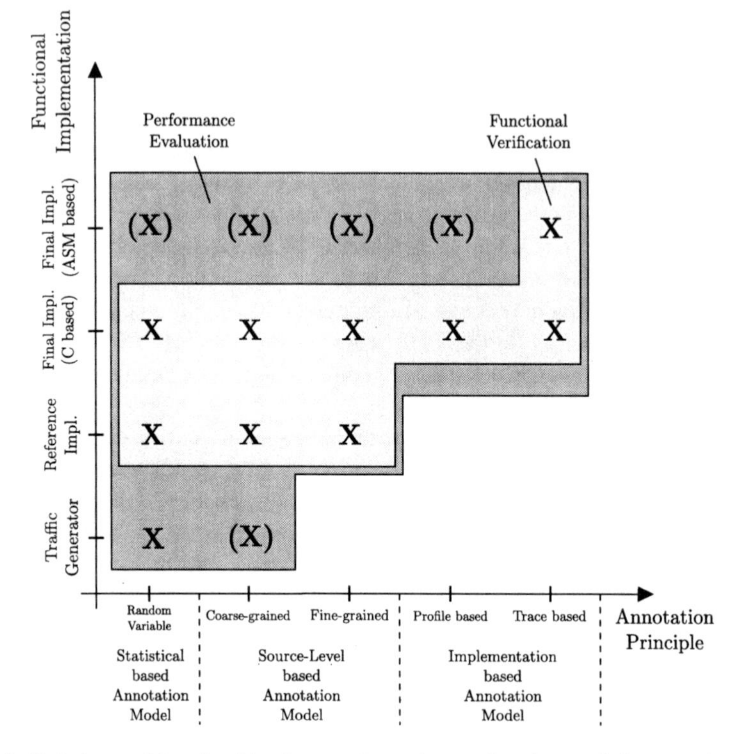

Fig. 7.3 Techniques of functional implementation and execution characteristic annotation

implementation is reached. Except when using traffic generators, the incorporated functional implementation supports verification whether the application executes correctly or not.

For efficient modeling, the execution characteristics (x-axis) and functional implementation (y-axis) have been kept orthogonal. When addressing functional verification, the VPU allows executing implementations that can run on the host processor core, commonly a x86 processor architecture. Typically, this requires a reference or final implementation in C/C++, whereas Assembly code for the targeted embedded processor cannot be executed. However, functional verification in the presence of a low level Assembly implementation can be done by using instruction set simulators. Undoubtedly, the use of traffic generators prohibits functional verification because no functionality is contained at all.

The key method for annotating a particular task's timing behavior is rather simple. At specific parts of the given task's source code, the execution characteristic of the code segment is passed to the VPU by a *consume(···)* statement. The function argument can be of arbitrary value ranging from a simple constant value to a complex C++ class which dynamically computes the annotation. Please note that the given basic annotation is in clock cycles allowing the VPU to easily reflect effects such as clock frequency modifications.

Definitely the quality of results, in terms of precise execution behavior, heavily depends on each component and its estimated execution characteristics. This demands different techniques for estimating/computing these annotations. Four of such techniques will be discussed in the following that serve as a foundation to cover a wide range of techniques, from early coarse-grained estimates down to trace-based annotations obtained by instruction set simulation measurements.

- *Statistical Annotation Model.* At the highest abstraction level and with only limited implementation knowledge developers can quickly apply statistical models. These models are typically based on rather coarse-grained estimates of the required operations and the underlying hardware architecture features. The Time Retrieval Engine (TRE) [267] can be utilized to ensure a smooth transition from the analytical implementation model to the abstract simulation model.
- *Source-Level Annotation Model.* This level of abstraction is characterized by a more fine-grained implementation knowledge. It provides a mechanism to annotate the execution characteristics within the individual tasks at arbitrary levels of detail. Hence, this abstraction level can be further subclassified based on the underlying implementation of functionality and timing annotations.

  - *Nonfunctional and Coarse-grained Annotations.* At early design stages the general question occurs how to implement a particular application given only as a pure mathematical algorithm. The lack of any functionally correct implementation at such design stages prohibits instruction set simulation. In contrast, this high abstraction level allows utilization of coarse-grained timing annotation on, e.g., basis of functions, neglecting the functionally correct implementation at first. However, such models can mimic the execution based on fine-grained estimates along with the communication demands based on communication requests in the style of traffic generators [268]. The natural refinement and implementation of the functionality lead to the next abstraction level.
  - *Functional and Fine-grained Annotations.* During implementation, constant update and refinement increases the precision of the individual task execution and, inherently, the precision of the obtained system-level results. This step-by-step development finally leads to a functionally correct implementation with a fine-grained annotation of the execution characteristic. As the VPU currently supposes implementation of the functionality in C and C++ this reference implementation can directly serve as foundation for the following refinement steps.

  Implementation-based Annotation Model. Having pieces or the complete functionally correct implementation for one processing element available, system architects are encouraged to utilize the implementation-based annotation model. When no processor core has yet been selected the annotation can be based on the μProfiler [255], whereas trace-based annotations are not applicable. With a particular processor core selected, trace-based annotations typically achieve better results and, hence, should be preferred.

- *μProfiler-based Annotations.* The μProfiler computes the timing annotations based on source-level performance estimation. This technique requires pure C source-code, while the instruction set architecture of the underlying processor core can still be in development stage. A fine-grained source-code instrumentation [170] builds the centerpiece of this technique to collect runtime statistics. This allows dynamically capturing the actual behavior during execution of the application. High precision in the accuracy has been reported for RISC-like architectures [269]. The precision is less accurate for irregular architectures, such as DSPs and ASIPs. Mostly, these architectures require the usage of low-level software constructs to exploit the full capacity of the underlying processor core [60] that cannot be captured by annotations based on the C programming language. This mismatch in terms of software programming can easily lead to predicted annotations diverging from the final implementation, even when the C-based approach operates well. In such cases system architects should consider selection of source-level or trace-based annotation methods.
- *Trace-based Annotations.* This last abstraction level before applying instruction set simulation, should be only considered as an intermediate step; hence, it should only be applied in special cases. Among them is the multithreaded software execution where some tasks/applications are fully implemented while others are lacking the implementation. Even missing a single task implementation prohibits the system-wide evaluation with an instruction set simulator. In contrast, the VPU model supports the mixing of different timing annotation principles to evaluate the system-wide performance prior to having the final software and hardware implementation. In addition, this annotation principle can be applied to increase simulation speed. However, latest instruction-accurate instruction set simulators, utilizing binary-to-binary translation, achieve extremely high simulation speeds. Therefore, system architects have to consider this on an individual basis.

These four major annotation techniques will now be discussed in detail starting from the highest abstraction level down to the lowest one.

7.3.1 Statistical Annotation Model

At the highest abstraction level the tasks are modeled as traffic generators that produce and consume data. Additionally, the task execution characteristic is reflected by a random variable because the implementation is not yet known and only rough performance estimates exist. Section 6.3 has already sketched the basic principle of refinement of the analytical implementation model to the simulation-based technique. For each task in the analytical model, a representative is constructed in the simulation-based environment. Such a task, apart from sources and sinks, is reflected by a simple traffic generator that consumes data, annotates the task execution characteristic, and generates data that is passed to the next task. The analytical model

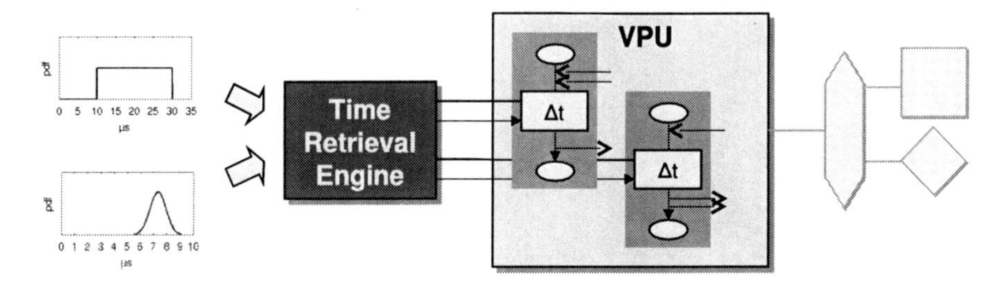

Fig. 7.4 Annotation of the execution characteristic: statistical model

considers each task execution behavior as a random variable $X(task, PE)$ with a probability density function (pdf). At simulation time a single value $\Delta t(task, PE)$ equal to the number of consumed cycles is generated by the use of the Time-Retrieval-Engine [267] (Fig. 7.4). This engine generates the numbers of cycles at random from the given pdf $X(task, PE)$ according to the inversion method, also known as the inverse transform sampling method (cf. [270, p. 27f.]).

Apart from these rather simple tasks, more complex ones can be constructed by utilizing typical control-level constructs of the C programming language, e.g., loops and if-then-else statements, as later shown within the source-level annotations. However, the underlying annotation technique remains equal. Based on a random variable for a specific code segment, the Time-Retrieval-Engine samples a single value, which – in turn – is utilized to annotate the behavior dynamically at run-time. This can be utilized to reflect fine- and coarse-grained task execution behaviors.

With ongoing development and definition of the architecture, the execution characteristics can be described in a more fine-grained manner, which allows lowering the abstraction level to source-level annotations.

7.3.2 Source-Level Annotation Model

This level covers a broad range of options, ranging from nonfunctional workload models with coarse-grained annotations to functionally correct tasks with fine-grained annotations (Fig. 7.3). When using coarse-grained annotations, instrumentation of tasks is typically performed per function or even per task, whereas fine-grained annotations are applied, e.g., per source-code line.

In order to demonstrate both techniques, Listings 7.1 and 7.2 highlight examples of their usage. The first case illustrates the annotation on the basis of instrumentation at the granularity of functions (Listing 7.1). The second example (Listing 7.2) highlights a more fine-grained instrumentation of the source code to keep track of the execution behavior. For RISC-like architectures typically distributed with compilers, such fine-grained C source-code instrumentation can capture the execution behavior well [250]. Source-level instrumentation is not necessarily imprecise when using an application specific processor core, like a DSP or an ASIP. However, these

```
1  MyTask::task()
2  {
3    ...
4    while (1) {
5      int values=10;
6      int result;
7      // per function_1 call
8      consume(values * EXEC_VALUE);
9      result = function_1(values);
10
11     // per function_2 call
12     consume(EXEC_PER_CALL);
13     result = function_2(result);
14   }
15 };
16
17 int MyTask::function_1(int values)
18 {
19   int result = 0;
20   int data;
21   for ( i = 0; i < values; i++ ) {
22
23
24
25     read(VALUES + i*size_of(int),
26          data, size_of(int) );
27
28
29     result += data;
30   }
31
32
33   return result;
34 };
35
36
37 void MyTask::function_2(int data)
38 {
39
40
41   data = data * data;
42   write(RESULT, data, size_of(int));
43 };
```

Listing 7.1 Coarse-Grained Annotations

```
1  MyTask::task()
2  {
3    ...
4    while (1) {
5      int values=10;
6      int result;
7
8
9      result = function_1(values);
10
11
12
13     result = function_2(result);
14   }
15 };
16
17 int MyTask::function_1(int values)
18 {
19   int result = 0;
20   int data;
21   for ( i = 0; i < values; i++ ) {
22     // per loop iteration
23     consume(EXEC_LOOP_VALUE);
24
25     read(VALUES + i*size_of(int),
26          data, size_of(int));
27
28     consume(EXEC_ADD_VALUE);
29     result += data;
30   }
31
32   consume(EXEC_RETURN_VALUE);
33   return result;
34 };
35
36
37 void MyTask::function_2(int data)
38 {
39   // per function_2 call
40   consume(EXEC_PER_MUL);
41   data = data * data;
42   write(RESULT, data, size_of(int));
43 };
```

Listing 7.2 Fine-Grained Annotations

architectures require a different software development than a standard RISC processor core. One reason is that standard C programming is mostly not suited as compilers cannot efficiently determine special instructions and cannot fully exploit the provided hardware features. Therefore, common software development for such cores relies on a mixture of high- and low-level software implementation to efficiently utilize the hardware architecture. Recent case studies have illustrated the performance gain between general purpose C and Assembly programming language to be easily in the range of one order of magnitude for a DSP-like architecture [60]. Hence, when targeting such architectures, fine-grained C-level instrumentation does not precisely model the low-level programming language. In turn this can be modeled fairly precisely at coarse-grained function level because developers typically know the exact execution characteristic of the software programmed in Assembly. These different methods for architecture and software modeling strengthens the need for different annotation techniques, which will be later illustrated within the case study (Chap. 8).

Following the design and refinement flow, the final output of the source-level annotation phase is a C-language-based reference implementation including annotation of the execution characteristic. Before entering the ISS-based ESL development, two intermediate abstraction levels can be applied to address special design issues. These implementation-based annotation levels and their methodology will be highlighted in the following.

7.3.3 Implementation-Based Annotation Model

Based on a reference or final software implementation, this approach automatically applies fine-grained execution characteristics. In the following, the two available techniques are introduced, which differ depending on whether the underlying processor architecture has already been selected or not. The first discussed approach operates on top of the μProfile [255] technology, which allows C-code profiling without the underlying hardware architecture being known or available at the current design stage. The later highlighted trace-based technology makes use of the targeted processor architecture and compiler in order to extract the software execution characteristic. This implies that an instruction set simulator and software toolchain are at hand.

μProfiler-Based Annotation Model

The key contribution of the μProfiler-based annotation [269] is to apply *fine-grained, automatic source-code instrumentation* techniques to implicitly model the execution characteristic as intratask memory accesses and estimating cycle counts. Figure 7.5 illustrates the general concept. The instrumenter (based on the μProfiler tool) inserts additional *instrumentation code* into the task source-code. Such an instrumentation code dynamically increments cycle counters and redirects the intratask memory accesses to the communication architecture when the corresponding task is executed on a VPU within a system level simulation.

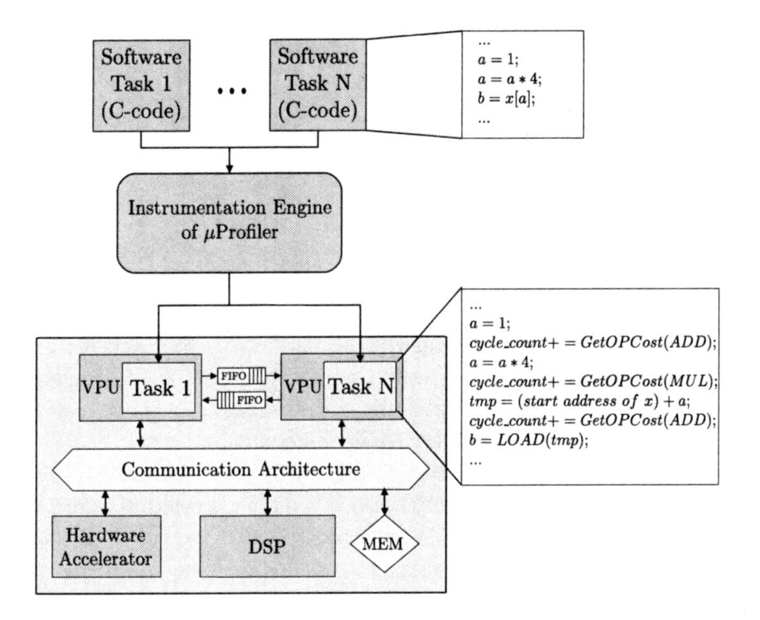

Fig. 7.5 Task execution work flow

The applied fine-grained software instrumentation is done on a *Three Address Code Intermediate Representation (3-AC IR)* level where all C operators and the majority of memory accesses are visible. Additionally, high-level standard IR optimizations, such as constant propagation, constant folding, and loop invariant code motion, can be performed on this IR. Such optimizations reduce chances of false prediction (such as counting operations that will be eventually eliminated by compiler optimizations) in estimating cycle counts and memory accesses. The accuracy of the software simulation is comparable to that of instruction set simulation for RISC-based architectures [250]. Later approaches have further extended such frameworks with features such as backends by utilizing LLVM [271,272] and have proved the vitality of such approaches in the domain of RISC processors.

In order to give a brief introduction, the overall principle will be discussed in the following. As all operators and a majority of the memory accesses are explicitly visible in 3-AC IR, the instrumenter, as shown in Fig. 7.5, only needs to add an extra C line after each IR operation to increase the cycle counter by the *operator cost* (obtained through *GetOpCost*). The system architect can assign each operator an appropriate cost that can be configured keeping the intended target processor in mind. For example, if the user intends to run a task on a processor which has a latency of three for multiplication and of one for addition then he can assign costs one and three to the addition and multiplication operators, respectively. The total estimated execution time of the task is given by the following formula:

$$\text{Cycles} = \Sigma_{i=1}^{n} E(O_i) \times C(O_i),$$

where $E(O_i)$ and $C(O_i)$ are the execution count and cost for an operator O_i, respectively.

The instrumenter also inserts function calls to intercept and report all accesses to different source level data elements, such as arrays, structures, and global variables, found in any application. Usually memory accesses, for an application written in a high-level language like C, originate from four sources:

1. Accesses to global *scalar and composite variables* (i.e., structures) and arrays.
2. Accesses to a function's local *composite* variables.
3. Accesses to *dynamically allocated memory* on the *heap*.
4. Accesses during building-up and cleaning-up of a function's stack frame.

The local scalar variables are usually allocated in registers, and the number of memory accesses caused by them is often negligible. The 3-AC IR makes the first three kinds of memory accesses explicit by converting all global accesses and local composite accesses to pointer *dereference* operations.

The simulation of intratask memory accesses (as done by the *LOAD* function in Fig. 7.5) is invoked with a *memory address*, and is expected to return the value contained in this memory address. At the same time, it is expected to simulate this memory access over the communication network and increment the cycle counter at the end of this simulation.

The technique of code instrumentation-based timing estimation and annotation has proved to work fairly well for RISC-like architectures. For more specific processor architectures like DSPs and ASIPs, this instrumentation technique typically lacks precision as it is based on the C programming language. In general this issue prohibits developers from exploiting the full performance capabilities of highly specialized architectures. Therefore, other techniques such as source-level annotations on the basis of functions can achieve more precise values and might be preferable.

When having a software implementation at hand and a processor architecture in mind, developers can apply the trace-based annotation technique. With respect to the targeted processor core this technique has the ability to achieve superior accuracy and will be introduced next.

Trace-Based Annotation Model

During ongoing development, parts of the complete system might be finalized with respect to the implementation of software and hardware. For example, a particular processor core has been selected and some tasks have already been implemented in software. Under the assumption of having a compiler and instruction set simulator available, system architects can easily simulate and measure the performance of these tasks. However, traditional performance evaluation of the complete system at the level of instruction set simulation would necessarily have to wait until the complete software and hardware have been implemented for the targeted architecture. To overcome this issue and to incorporate the most precise implementation knowledge at early design stages, trace-based annotation is the optimal vehicle to overcome such late performance investigations. The fundamental idea is to later incorporate the measured performance characteristics into a VPU-based simulation. Moreover, in contrast to the approach discussed earlier, this method can be applied when using high- or low-level programming languages.

Overall, the annotation of the execution characteristic is separated into two phases (Fig. 7.6). In the first phase the already available software is executed on the instruction set simulator or within a given subsystem. During this execution, events are recorded that occur at the borders of the investigated core/subsystem. Typically, these are caused by memory accesses and interrupts. It should be noted that instructions set simulators do not normally provide such tracing facilities, hence developers have to insert them manually.

In the second phase, the corresponding task running on the VPU reads the traced events and replays them to mimic the exact behavior of the measured software part in a later system simulation. The mixing of arbitrary annotation principles is the intended use case to allow performance investigations before the final implementation is ready. This allows characterization of each task with the most precise implementation knowledge available.

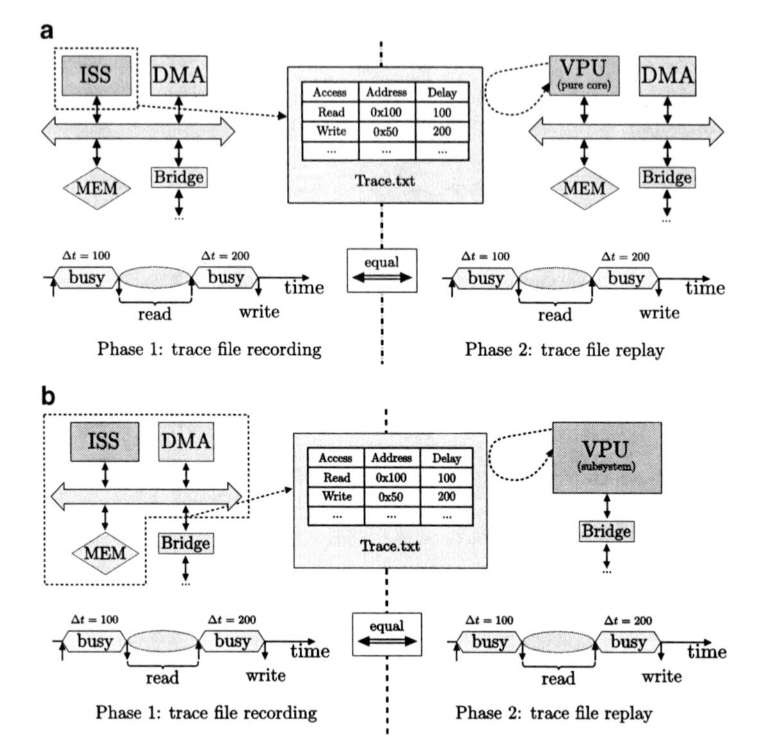

Fig. 7.6 Annotation of the execution characteristic: trace-based annotation. (**a**) Trace-based annotation for processor core. (**b**) Trace-base annotation for subsystem

A fundamental part of all annotation methods is the accuracy that is achieved at each level. Hence, this work dedicates one single section (Sect. 8.1) discussing the achievable accuracy. When lowering the abstraction level toward the final implementation, the next level is naturally instruction set simulation. For acceptance and practical use, a continuous refinement methodology from high abstraction level to ISS-based level is mandatory. The VPU technology naturally supports such a smooth transition as is highlighted in Sect. 7.6.

Today's software development typically makes use of software layers and libraries, such as device drivers, middlewares, and Operating Systems (OSs). Inspired by the traditional ISO/OSI-model [11] and OS layers [273], the VPU model similarly incorporates the most common layers. In addition, simple enhancement and extension of such layers can be applied because of the modular structure of the VPU model. Since software layers heavily impact the overall execution, these can be annotated to capture the execution behavior. In the subsequent sections the software layers integrated on top of the VPU are highlighted.

7.4 Software Layers of the VPU

The subsequent discussion of the VPU's software layers introduces them in a bottom-up fashion starting with the lowest layer which is the hardware abstraction layer. Later, the layers of the device drivers, the operating system, and the middleware are highlighted.

7.4.1 Hardware Abstraction Layer

The large variety of different hardware platforms in the domain of wireless communication makes portable software mandatory to reduce development costs for next-generation devices. Portability demands an abstraction that separates software development from the hardware-specific and dedicated features of each particular platform. Similar issues in the domain of general-purpose computing resulted in the application of a thin software layer called *hardware abstraction layer* (HAL), e.g., within the Windows NT and later operating systems. This abstraction layer completely separates high-level software, such as the operating system and device drivers, from the underlying hardware platform by means of representing abstracted devices to the rest of the software. Typically, software can access such devices via routines in order to simplify software development on top. This enables an easy porting of the software from one platform to another, as merely the implementation of the HAL needs to be modified. All other software should be portable by utilizing a different compiler toolchain.

In general, the HAL supports services like platform-independent device accessing, interrupt handling, DMA transfer management, and low-level operating system kernel operations such as context switching and synchronization services. While the HAL provides the fundamental services, that are platform specific, device drivers like display controllers and keyboard devices are *not* included and their development is on top of the HAL in order to conserve portability.

Similar to common hardware abstraction layers, the HAL execution needs to reflect the arbitrary hardware features of various processor cores. As data transfer has long been a key challenge in processor core design, the range of available communication features of modern processor cores is vast. Starting from a simple single shared-bus port, modern processor cores nowadays incorporate ports for dedicated point-to-point communication, multiple ports connected to different shared buses and/or even complex Networks-on-Chips. For example, Tensilica's configurable Xtensa LX2 [48] processor core can be equipped with additional general-purpose I/O (GPIO) and FIFO interfaces, in addition to the standard bus interfaces. To reflect such arbitrary interfaces, the VPU is kept generic so that it can be equipped with an arbitrary number and type of communication ports to mimic the various processor cores.

With respect to the final hardware architecture, or a virtual platform operating on instruction set simulation, the HAL needs to be implemented in Assembly

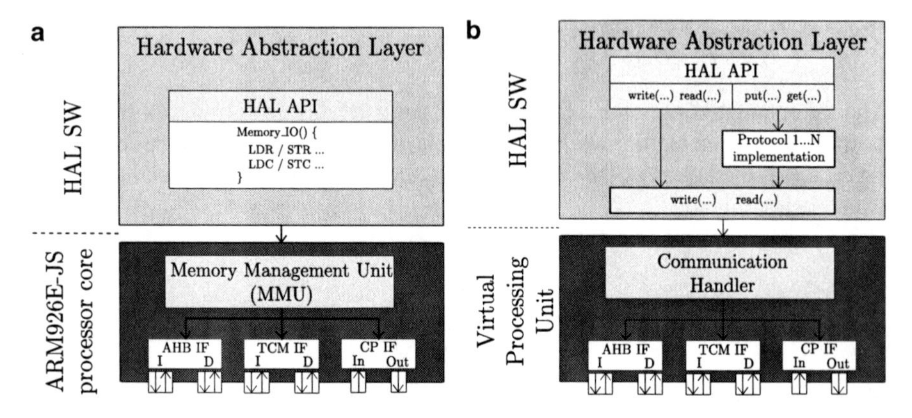

Fig. 7.7 Comparison between hardware abstraction layer for ISS- and VPU-based simulation. (**a**) Hardware Abstraction Layer for ARM926E-JS. (**b**) Hardware Abstraction Layer on VPU

programming language or low-level software that holds hardware specific information. Typically load/store operations are handled based on the principle of memory-mapped I/O. Based on the address, the memory management unit (MMU) determines which communication port of the underlying processor core deals with the communication request. Other more specific communication architectures can be addressed by special instructions such as ARM's coprocessor interface [174] or the Fast Simplex Links (FSL) of Xilinx' Microblaze processor [274]. Figure 7.7a illustrates the execution of a HAL on an ARM926E-JS [174] processor core. While standard load (LDR) and store (STR) register operations use the memory address, the special instructions load (LDC) and store (STC) coprocessor register are directly passed to the coprocessor interface.

Instead of using hardware-specific software code as input, the VPU's HAL provides a highly generic communication interface to emulate all kinds of communication accesses. It is based on the common principle of data communication, i.e., the sending and receiving of data. Therefore, the HAL provides the two routines *put* and *get* to enable efficient communication accesses. During simulation these communication requests are either directly passed to the communication handler of the VPU or trigger a protocol execution. Figure 7.7b highlights the basic principle of the VPU while it mimics an ARM926E-JS processor core. Direct communication accesses (*read* and *write*) execute similarly to the behavior of the real hardware. Depending on the memory address, the access is routed to the right communication port. As these direct communication accesses are platform specific, developers should utilize communication protocols that are built on top of the direct communication interface. These protocols can vary from a simple access of a dedicated memory address, e.g., reading the current clock value, to larger protocol implementations that utilize DMA controllers to copy a complete block of data. In order to access external processor ports these implementations utilize the direct communication interface.

Table 7.1 Protocols of the VPU's hardware abstraction layer

Data structure	Communication port access	Special features
Variable structure (C-based)	Internal register	Blocking/Nonblocking
	External memory mapped	No/Flag-based synchronization
	Specialized communication port	
	DMA transfer (memory mapped)	

Implementations of such protocols are not limited by any restriction in general. Nevertheless, being located within the hardware abstraction layer, these protocols should focus on rather low-level memory accesses rather than complex device drivers. Protocol implementations differ in the transferred data structure, the accessed communication port, and other features such as flag-based synchronization or blocking and nonblocking calls. Table 7.1 gives an overview of exemplary protocols whereas the pseudo-code snippet in Listing 7.3 shows a nonblocking and nonsynchronized DMA protocol implementation.

With the hardware abstraction layer deployed on top of the VPU simulation model, arbitrary software can be implemented without the need to incorporate hardware specific knowledge of memory addresses and other features. After introducing the low-level hardware abstraction layer, the discussion now turns to more software-centric layers like the device drivers and the operating system layer of the VPU.

```
DMA_Protocol() {
  if (write) {
   // trigger DMA write
   // Clear pending IRQs
   write(DMA_reset_address, 0, size_of(int));
   // Set DMA transfer size
   write(DMA_set_size_address, size_of(data), size_of(int));
   // SRC & DST Address
   write(DMA_set_src_address, src_address, size_of(int));
   write(DMA_set_dst_address, dst_address, size_of(int));
   ...
   // Enable DMA transfer
   write(DMA_enable_address, 1, size_of(int));
   wait_for_irq();
  }
  else {
   // trigger DMA read
   ...
   wait_for_irq();
  }
}
```

Listing 7.3 Exemplary DMA Protocol based on CoWare's generic DMA controller [158]

7.4.2 Device Drivers

Implementation of wireless communication devices follows the principle of component-based design, like the TI OMAP platform visualized in Fig. 2.4. The fundamental idea of such design is to assemble IP components of different type and arbitrary number. Advantage is the fast and simple platform development, but well-defined component interfaces are a *must*. The centerpiece of each platform

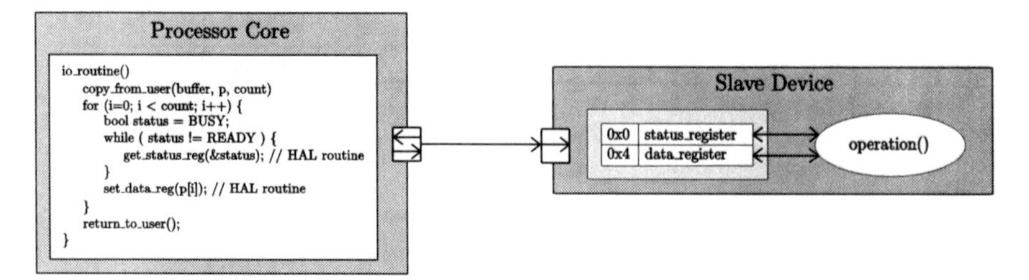

Fig. 7.8 Hardware device and device driver [273, p. 285] with pure slave behavior

are the processor cores that execute the user applications and perform platform management. Nevertheless, other processing elements are mandatory to achieve the application-given requirements. These hardware components range from simple I/O controllers, like a keyboard controller, to highly complex programmable 2D/3D graphic hardware accelerators [41].

From the software perspective, these hardware components are commonly named *devices*. As component-based design mandates well-defined interfaces in hardware, they are naturally reflected by *device drivers* in the software domain. These drivers ease software development as they abstract low-level from high-level software development and hide all hardware-specific device information. In general, device drivers provide various access routines to software developers, which range from simple memory-mapped I/O components to complex access patterns based on DMA data communication. In addition, interrupt handling is of huge importance to prevent active waiting times due to polling operations.

Tanenbaum [273, p. 284ff.] distinguishes three basic types of devices, namely *programmed I/O*, *interrupt-driven I/O*, and *I/O using DMA*. The simplest form, i.e., the programmed I/O device, is exemplified in Fig. 7.8. Here the hardware device exposes an interface consisting of two registers to the rest of the system. While the first register denotes the current status of the device and the second enables acceptance of arbitrary data. For classical slave devices, the arrival of new data triggers execution of the functionality, e.g., displaying of the received data on an LCD display.

The illustrated software routine of the programmed I/O device driver (Fig. 7.8) polls until the status flag signals a nonbusy device before passing new data to it. This mechanism is called polling and has the disadvantage that the processor core is always active when waiting for the device to be accessible. Hence, more elegant device drivers make use of advanced features such as interrupt-driven I/O, or completely outsource the data transfer to a direct memory access (DMA) controller. Such implementations enable the processor core to execute other tasks or to enter sleep mode until the device is available. Appendix B.1 highlights examples of two more advanced device drivers and their VPU implementations.

With respect to this example, the depicted pseudo-code implementation highlights the access to the underlying device by calling HAL routines (Listing 7.4). Similar to these routines, the HAL (Listing 7.5) allows efficient implementation of

```
io_routine() {
    copy_from_user(buffer, p, count)
    for (i=0; i < count; i++) {
        bool status = BUSY;

        while ( status != READY ) {

            // HAL routine follows
            get_status_reg(&status);

        }


        // HAL routine follows
        set_data_reg(p[i]);
    }
    return_to_user();
}
```

Listing 7.4 Example of polling memory mapped I/O device driver [273]

```
io_routine() {
    copy_from_user(buffer, p, count)
    for (i=0; i < count; i++) {
        bool status = BUSY;
        consume(...);
        while ( status != READY ) {
            bytes = size_of(bool);
            // HAL routine follows
            read(STATUS_ADDR, &status, bytes);
            consume(...);
        }
        consume(...);
        bytes = size_of(p[i]);
        // HAL routine follows
        write(DATA_ADDR, p[i], bytes);
    }
    return;
}
```

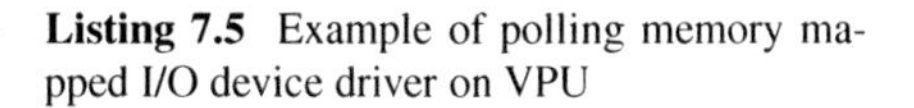

Listing 7.5 Example of polling memory mapped I/O device driver on VPU

device drivers on top of the direct-interface API (write- and read) and other protocol implementations. This enables the implementation of device drivers that are equal to the final implementation for a particular hardware.

In order to support early design space exploration the VPU technology operates on top of host-based simulation with an annotation principle of the execution characteristic. When using instruction set simulation or the real hardware, developers can measure the execution characteristic of the cross-compiled software. Unfortunately, neither the instruction set simulator nor the hardware is available in early design phases; also the compiler tool-chain is not in its final stage. Hence, as previously highlighted (Sect. 7.3), the VPU provides a unique methodology to annotate execution characteristics on different abstraction levels to mimic the behavior of the real hardware. Selection of the utilized abstraction layer in order to best capture the behavior, depends on the expert knowledge of the designer. Therefore, no restrictions occur for instrumentation of device drivers so as to annotate their execution characteristics. Listing 7.5 exemplifies a source-level-based annotation scheme (Sect. 7.3.2).

With the already introduced software layers, processor cores like ones with a single-threaded application (Fig. 7.2) or programmable hardware accelerators (Fig. 7.2) can be captured by the VPU. With operating systems entering embedded systems, in particular for wireless communication devices (already 80% of developed embedded systems use an operating system in 2005 [275]), the next logical extension of the VPU enables support for operating systems and other middlewares. First, the foundation of operating system modeling, and later of middleware modeling is introduced.

7.4.3 Operating System Layer

Finding a unique and precise definition of an operating system is rather difficult, as the functionality and purpose of the operating system heavily depend on the point of view. For example, in a top-down software-centric view, the operating system is an extended machine that shields software developers from low-level hardware issues. Taking a bottom-up hardware-centric point of view, the operating system exhibits a central resource manager of the available hardware resources.

Instead of selecting one exclusive definition, today's common agreement is to subsume both definitions into a single one that defines the basic features of an operating system as a:

1. Management and coordination of resources and activities (applications).
2. Acting as an extended/virtual machine to relieve application software developers from hardware details by inserting an abstraction layer above the hardware (Fig. 7.2).

Research in operating systems has quite some history that can only be sketched here. More in-depth discussions can be found in [273]. Basic research started in the 1960s with the introduction of batch systems. While in following years technology advances increased the computational performance of computers, I/O-access capabilities were lagging behind. This resulted in long stall times of the computational resources due to the waiting for data. As a solution, the paradigm of multitasking was invented and applied to overcome this issue. Later this evolved to one of the absolutely fundamental principles of modern operating systems for resource management. With the introduction of very large scale integrated (VLSI) circuits, the capabilities of computational resources exploded (see Moore's Law [3]), which – in turn – boosted software development and, in particular, the use of operating systems.

The various different objectives and business fields resulted in a vast number and type of operating systems. The general-purpose domain as we see it today is dominated by Windows and Unix (including derivatives such as Linux [276]) operating systems. Nevertheless, for each particular domain a specialized and dedicated operating system exists. Separation is commonly done on the basis of the addressed platform [273, p. 19f] which are mainframe, server, personal computer, embedded system, and smart card operating systems.

For the addressed application domain, focus is placed on operating systems for embedded systems. In contrast to operating systems from the personal computer domain, these are specially tailored for devices with limited resources such as restricted computational performance of the underlying hardware platform, memory size, as well as tight energy and power restrictions. In addition, embedded systems often require real-time capabilities leading to real-time operating systems (RTOS).

The market of operating systems for embedded devices is widely distributed among various commercial vendors, such as Windows CE, VxWorks, QNX, RTX, C/OS, and Mentor Graphics' Nucleus RTOS, as well as open-source solutions built upon Unix derivatives like Linux [275]. The common features such as process management and control, as well as interprocess communication and synchronization, are implemented in every operating system. Other more dedicated features and their implementation depend on the utilized operating system.

Despite the aforementioned commonalities, operating systems vary in features, implementations, and API usage so that software portability is limited and mostly requires manual modification of the software. In order to make software portable, an interface standardization called Portable Operating System Interface (POSIX) [277] was defined. POSIX became the IEEE Standard 1003.1-1988 that defines a fixed API on top of UNIX. It can also be utilized in any other operating system. Fundamental features supported by the standard are process control and creation, as well as the handling of I/O and other data communication. Later additions to the original standard include real-time support along with advanced scheduling and synchronization principles.

Returning to the discussion of the VPU technology and the operating-system layer, early design space exploration aims at the identification of necessary features and evaluating which operating system serves best for the addressed application and platform. In addition, all effects of the operating system, such as context switches, can have significant impact on the overall system performance. Therefore, this must be taken into account right from the start of the design cycle. The modeled and available operating-system layer supports the fundamental concepts of:

- Process control and management.
- I/O and interprocess communication.

To capture the effects of resource management during early design space exploration, a generic operating system (generic OS) has been developed on top of the VPU. This generic OS does not reflect a single, unique operating system. Instead, it incorporates the identified common features. This allows software developers to evaluate the necessary functions to achieve a successful design. The abstract simulation-based investigation allows the evaluation of arbitrary design decisions and replaces an ad hoc selection of an operating system as commonly done today [275].

The features included in the generic OS have been identified during a detailed inspection of two candidates, namely the IEEE POSIX standard and the Real-Time Operating System for Multiprocessor Systems (RTEMS). These features are:

- Process management:
  - Multitasking including, e.g., dynamic task creation, starting and stopping of tasks.
  - Preemptive and nonpreemptive scheduling with various scheduling algorithms, e.g., round-robin, priority based, or time slicing [261].
  - Advanced process state control (on basis of a standard control as in [273, p.78f]).
- Interprocess communication and synchronization:
  - Synchronization primitives like mutexes and semaphores [273, p. 110ff].
  - Device drivers and I/O interfacing

Resource management, including process management, is the centerpiece of each operating system. Therefore, the generic OS is formed around the common principle of multitasking. An advanced task-state control was selected to capture arbitrary configurations as highlighted in Fig. 7.9. In total, the proposed framework addresses

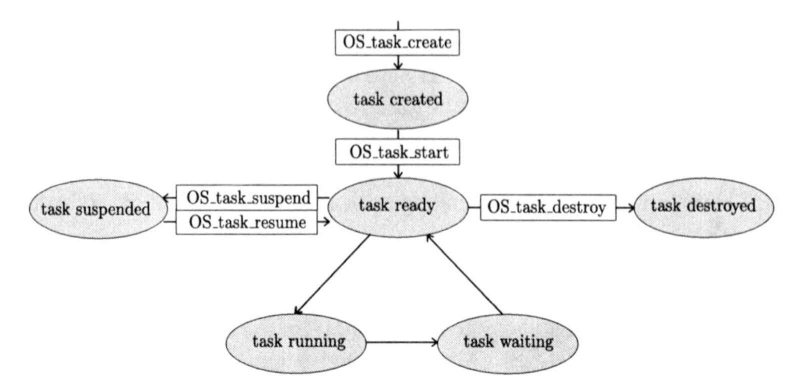

Fig. 7.9 Advanced task state control in the generic operating system

the objective of design space exploration. Hence, in contrast to emulation of one unique operating system, the generic OS provides a set of fundamental functions in order to support exploration of required features. Table 7.2 depicts the common features that have been identified and incorporated for task management. For comparison, Table 7.3 illustrates the correspondence between the generic OS and the reference implementations of the RTEMS and the POSIX API.

A typical design flow starts with the use of these generic functions. During later design stages, additional required features that need to be incorporated might be identified. Therefore, the generic OS is structured as a software layer above the VPU simulation model, which allows easy addition of arbitrary extensions. With the ongoing development process, the level of abstract simulation is left and the level of instruction set simulation is entered according to the envisioned design process. Naturally, at this level the software developed for the generic OS needs to be ported to the specific OS in a (semi-)automatically or manual process. These refinement steps are discussed in depth later within Sect. 7.6.

Table 7.2 Basic functions of the generic operating system to support task management

OS_Task_Create	Creates a task including allocation of memory.
OS_Task_Destroy	Destroys a task and releases its data.
OS_Task_Start	Starts a task and enables the scheduler to trigger the task execution.
OS_Task_Stop	Stops a task and removes it from scheduling.
OS_Task_Suspend	Suspends a task and removes it from scheduling.
OS_Task_Resumes	Resumes an earlier suspended a task.
OS_Task_Wait	Yields the task till the next time the scheduler initiates task execution.
OS_Task_Wait_event	Yields the task till the event is activated.
OS_Task_Wait_for_time	Yields the task and allows scheduling only after the given time.
OS_RoundRobin_Scheduler	Instantiates a round-robin scheduling method.
OS_Priority_Scheduler	Instantiates a priority based scheduling method.
OS_TDMA_Scheduler	Instantiates a time slicing scheduling method.

Table 7.3 Specific OS API refinement of important OS functions

Generic OS API	RTEMS API	POSIX API
OsTaskId	rtems_task_ident	getpid
OsTaskSuspend	rtems_task_suspend	pthread_kill(..., SIGSTOP)
OsTaskResume	rtems_task_resume	pthread_kill(..., SIGCONT)
OsTaskYield	rtems_task_suspend	sched_yield
OsTaskYieldTo	rtems_task_resume	pthread_kill(..., SIGCONT)
	rtems_task_suspend	pthread_kill(..., SIGSTOP)
OsTaskCreate	rtems_task_create	pthread_create
OsTaskDelete	rtems_task_delete	pthread_exit
OsTaskStart	rtems_task_start	–
OsWakeAfter	rtems_task_wake_after	sleep
OsShutdown	rtems_shutdown_executive	exit
...	...	...

Interprocess communication and synchronization forms the second basic feature of the generic OS. From the perspective of software development, interprocess communication operates on software constructs that ensure software communication and synchronization. Example constructs are semaphores, mutexes, pipes, message queues [273], and other more advanced ones like CORBA [278] or the message passing interface (MPI) [279]. This software-centric view is dominated by the provided mechanism and functionality.

Inspecting interprocess communication from the hardware perspective, each communication and synchronization represents a particular functionality in software that finally maps to a specific data structure stored within the memory of the hardware platform. Hence, all functions finally map to memory-mapped I/O and/or to the hardware abstraction layer interface. The fundamental goal of the VPU technology is to support early design space exploration. Consequently, the more fine-grained hardware perspective is chosen as it incorporates the complete software and hardware issues, while the software view abstracts the hardware to simplify software development.

The interprocess communication and synchronization is captured similarly to device drivers. Each primitive maps to a data structure and a set of functions to interface this data. Listing 7.6 in Fig. 7.10 illustrates a semaphore that is defined as data structure given by a counter variable and queue. The semaphore implementation comes with an initialize, a P- (prolaag-) and V- (verhoog-) operation [273, p. 110ff] and the data structure is stored as bits and bytes in any kind of memory.

The difference between a cross-compiled software executing on an instruction set simulator, and a VPU-based implementation of a semaphore is exemplified in Listing 7.6 and 7.7 in Fig. 7.10. Generating the software binary from a standard C-code implementation, the linker allocates a particular memory segment for the given semaphore data structure. This segment is mapped to an arbitrary memory address that allows physical access in order to support the required communication or synchronization. In contrast to instruction set simulation that executes the binary executable step by step, the VPU technology simulates the software on the

```
struct Semaphore {
   int count;
   Queue queue;
};

// Linker decides memory address
struct Semaphore s;
OS_event e;

void initialize (struct Semaphore s, int number) {

   s.count = number;

   s.queue.empty();
}

void acquire (struct Semaphore s) { // P-operation


   if (s.count <= 0) {
      OS_task_wait_event(e);
   }

   s.count--;
}

void release (struct Semaphore s) { // V-operation


   s.count++;

   if (s.count > 0) {
      e.notify();
   }
}
```

Listing 7.6 ISS

```
Semaphore s {
   int count;
   Queue queue;
};

// Sets explicitly the memory address
Semaphore *s = SEMAPHORE_MEM_ADDR; OS_event e;

void initialize (struct Semaphore *s, int number) {
   // write initial count value
   write(s+0, number, size_of(int));
   // clear queue
   write(s+size_of(int), 0, size_of(queue));
}

void acquire (struct Semaphore *s) { // P-operation
   int tmp=0;
   read(s+0, tmp, size_of(int));
   if (tmp <= 0) {
      OS_task_wait_event(e);
   }
   tmp--;
   write(s+0, tmp, size_of(int));
}

void release (struct Semaphore *s) { // V-operation
   int tmp;
   write(s+0, tmp, size_of(int));
   tmp++;
   read(s+0, tmp, size_of(int));
   if (tmp > 0) {
      e.notify();
   }
}
```

Listing 7.7 VPU (no annotation statements)

Fig. 7.10 Semaphore ISS vs. VPU software code comparison

host machine. This makes cross-compilation not required and allows system wide performance evaluation before the final hardware, compiler, and/or software implementation is available. However, this requires making memory accesses *explicit* within the software implementation of such communication and synchronization primitives. Otherwise, during simulation, the variable is treated as internal and no external memory accesses will be generated by the VPU. This does not necessarily lead to erroneous implementations, but might result in too optimistic performance results, as cost intensive memory accesses might be missed.

Listing 7.7 in Fig. 7.10 illustrates an implementation with explicit memory access based on the semaphore example. Here all accesses to the data structure are made explicit by utilizing the memory access routines of the hardware abstraction level (Sect. 7.4.1). For task management, the above introduced operating-system resource-management facilities are utilized and, for simplification reasons, annotations are neglected.

The implementation of communication and synchronization services closely relates to the earlier discussed protocol implementations of the communication primitives (put and get) within the hardware abstraction layer. In principle, one implementation method can replace the other, as both can utilize the functions provided by the operating system. Nevertheless, for the sake of a clear software-layer structure, this should be and is prevented for the VPU technology. Hence, low-level routines are kept in the hardware abstraction layer, minimizing the used operating-system features. Functions that extensively require the operating system for implementation, like semaphores, are kept within the generic OS layer to prevent the mixing of different software abstraction layers. Later this clear separation significantly simplifies the porting of the software application from the generic to a specific operating system.

With the synchronization and communication routines available, the generic OS support already provides a solid foundation for the inspection of operating-system features at early design stage. This includes capturing arbitrary performance effects and the impact of the operating system on the overall performance. Before introducing the implementation of tasks from the user perspective, the final layer subsuming other middlewares should be briefly sketched.

7.4.4 Middleware Layer

The name middleware originates from the domain of distributed systems. It defines a software that allows other software components and/or applications to connect, in the sense of data exchange, when running on different systems attached to an arbitrary interconnect network. Placed between the application software and the underlying operating system layer, it is called middleware. Whether a function belongs to the operating system or to the middleware layer is to some extend arbitrary. However, the general rule is that kernel functions belong to the operating system while the middleware incorporates extensions mostly targeting efficient interprocess communication. This rough and floating separation has the potential that over time original middleware features migrate into the operating system, as for example of the TCP/IP stack.

Because of the increasing usage of heterogeneous MPSoC platforms, middleware is becoming rapidly more important in the domain of embedded systems. Originally intended for distributed systems, middlewares have a rather large memory footprint and require high computational performance. Therefore, recent research has focused on middlewares for resource-limited embedded systems. Examples are specially tailored middlewares like the CORBA ORB express [280].

As a consequence, the VPU technology also has to provide suitable middlewares. Additionally, it must allow easy extension and development of middlewares to support evaluation of the corresponding performance impact at early design stage. The general layer concept allows for a simple development of middlewares on top of the generic OS layer and the VPU. In the following this will be sketched on the basis of the example depicted in Fig. 7.11.

The example shows two tasks, a source and sink task, executed in sequence. The source task (task #1) generates data tokens which are passed to the sink task (task #2). The sink task consumes all data. In addition, each execution of the sink task is independent of the other executions of that task, hence parallelism can simply be added by multiple instantiation. For the given example, the source task (task #1) generates the data tokens every $1000tu$ which is shorter than the sink task (task #2) requires for execution ($1000..2000tu$). Therefore, multiple instances of the sink task are necessary in order to capture each generated data token.

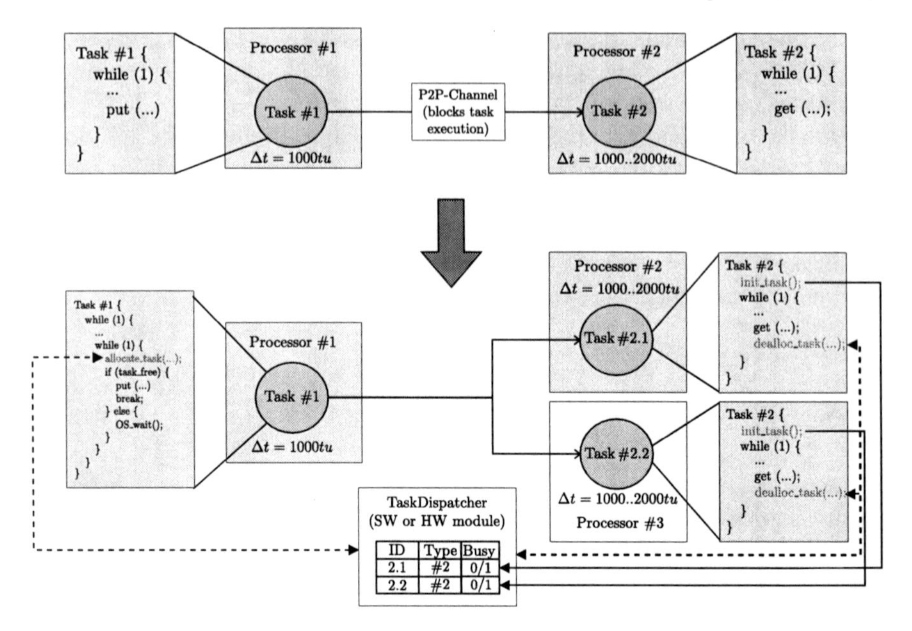

Fig. 7.11 Example of a middleware on top of the VPU

When instantiating multiple instances of the sink task, the source task has to decide whether to send the data token to the first or the second instance. One option is to utilize a static scheduling, e.g., a round-robin scheduling which decides where to send the data token. As long as the sink task runs on the processing element exclusively, this typically works well. However, the more likely case, especially when using processor cores, is that the resource is shared among multiple tasks so that the source task can never be sure whether the sink task has finished its prior execution.

In order to select the available task instance, the *task dispatcher module* and its *middleware* provides an efficient mechanism to overcome this issue (Fig. 7.11). At instantiation time, each task with multiple instances registers at the task dispatcher module. During execution, whenever the source task wants to send its data, it first requests and allocates an instance of the sink task by accessing the task dispatcher module. When getting a successful lock on a task instance, the data is sent to the corresponding task. Otherwise the source task must wait until a task instance is available. Data reception triggers execution of the corresponding sink task, which at the end signals the task dispatcher module the deallocation before entering the wait state. This releases the lock of the task instance and a following allocation request of the source task can be answered with this task instance.

Obviously, the middleware development does not affect any of the underlying software layers of the VPU so that the development can be performed independently. Having completed the discussion of all available and extensible software layers as well as the VPU simulation model, the subsequent section inspects the application development.

7.5 Application Layer

The proposed framework offers two design entries for modeling user applications. Besides a traditional textual design entry for software development, an efficient graphical design entry is provided. The graphical entry addresses, in particular, applications which can be consistently described in process networks, like synchronous dataflow graphs (SDFs) and Kahn process networks (KPNs). This is particularly well-suited for applications from the wireless communication domain. After highlighting the fundamental textual task modeling, the graphical design entry will be described.

7.5.1 Textual Design Entry

Traditional software development for an embedded system focuses on task-level modeling in a high-level programming language, typically C/C++. In addition, software developers make extensive use of libraries to increase reuse and fasten the development process. Examples of such libraries are operating systems, but also other device drivers and middleware libraries.

To support such development, the VPU technology incorporates the commonly used libraries and features as sketched earlier. Accordingly, software development at this abstraction level equals traditional development and enables a smooth (semi-)automatic refinement in order to achieve the final implementation.

To put the task-level modeling onto a solid mathematical foundation, an operational semantic based on the tagged signal modeling (TSM) [81] has been derived for both the task modeling and the underlying VPU simulation model [250]. This formal description defines a task execution as a *timed communication extended finite state machine* (tCEFSM) to capture arbitrary software execution on a processor core. After introducing the operational semantic, practical considerations of the textual design entry follow. Here, the actual task-level development in C/C++ programming language will be introduced and the correspondence between the practical and formal description is highlighted.

Operational Semantic

The theory of the tagged signal modeling (TSM) and timed communication extended finite state machines (tCEFSM) serves as foundation for the operational semantic of the VPU technology, including the task modeling in general [250]. After some basic definitions the task modeling itself is derived.

Elementary Definitions

According to tagged signal modeling [81] an *event* e consists of a time tag $t \in \mathcal{T}$ and a value $v \in \mathcal{V}$. With respect to general programming concepts, the value v represents an arbitrary abstract data type (ADT). Overall, such abstract data type can consist of various data fields including, e.g., address and data value of a memory request. The point operator serves to access a particular member of the data structure, i.e., $e_i.value.address$ denotes the address field of the event e_i. Based on the fundamental definition of a single event, a *signal* s defines a set of events, which can be viewed as a subset of $\mathscr{T} \times \mathscr{V}$.

Timed Communication Extended Finite State Machine

A software task executed on a particular processor core can be considered as a timed Communication Extended Finite State Machine (tCEFSM), which can be formally described. A tCEFSM originally derives from a finite state machine (FSM) [281]. To be more generic it is extended with internal variables (extended FSM) and output communication (communication extended FSM).

Definition 7.1 (Timed Communication Extended Finite State Machine (tCEFSM)). A tCEFSM defines an 8-tuple $(\mathscr{I}, \mathscr{O}, \mathscr{Z}, z_0, f, \mathscr{U}, \mathscr{D}_{busy}, \mathscr{D}_{delay})$ with:

- A set of input events $\mathscr{I} \subseteq \mathscr{T} \times \mathscr{V}$, and output events $\mathscr{O} \subseteq \mathscr{T} \times \mathscr{V}$,
- A finite, nonempty set of explicit states $\mathscr{Z}$,
- An initial state z_0,
- A set of variables $\mathscr{U} = (u_1, \ldots)$, which represent the implicit state,
- A state transition function $f : \mathscr{Z}^* \times \mathscr{I} \mapsto \mathscr{Z}^* \times \mathscr{O}$, where $\mathscr{Z}^*$ denotes the set of all implicit and explicit states,
- A set of busy periods $\mathscr{D}_{busy} = \{\Delta t_{busy,i}\}$, and
- A set of processing delays $\mathscr{D}_{delay} = \{\Delta t_{i,d}\}$.

The tCEFSM formally specifies the task modeling and allows defining the VPU model formally as done in [250] and [100] in respect to the architect's view framework (AVF) [282]. For practical considerations, software developers rely on programming languages like C/C++ [283, 284] rather than sticking to such formal finite state machines. Hence, in the following, the practical considerations highlight task modeling from programming perspective and link the utilized constructs to the formal definition.

Practical Considerations for Task Modeling

The formalized operational semantic defines a solid mathematical foundation of task modeling. However, software development in general follows a more practical and pragmatic design process rather than a strict formal approach.

Software development for embedded systems is typically carried out in a high-level programming language, such as C/C++. Therefore, it is of vital importance to efficiently support these programming languages in the abstract implementation model. However, there are inevitable minor differences between standard software development based on cross-compilation and the use of Virtual Platforms (VPs) [285] on the one side, and VPU-based simulation on the other.

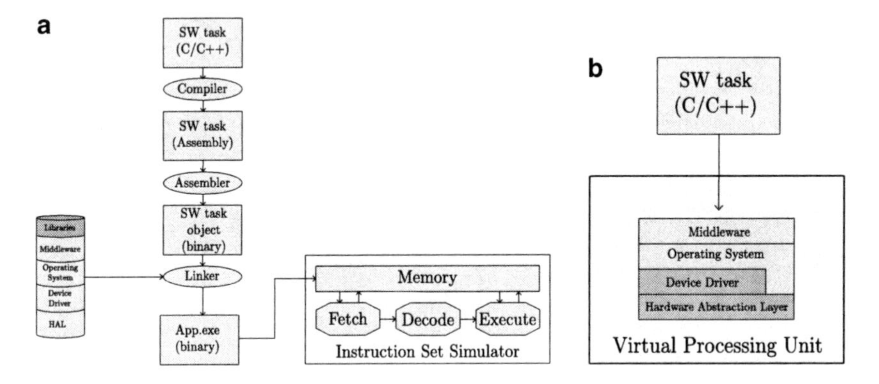

Fig. 7.12 Comparison of modeling of software and hardware on ISS and VPU. (**a**) Modeling and usage of software on an ISS. (**b**) Modeling and usage of software on a VPU

Figure 7.12 compares traditional instruction set simulation against the VPU technology. In addition, the corresponding software development flow is highlighted. Instruction set simulation mimics a processor core by reading the target executable instruction by instruction. During execution these instructions are decoded and the simulator emulates the intended operation of the addressed instruction set architecture. This definitely requires a software cross-compilation for the targeted architecture because the simulator operates on the target executable. This executable in binary format is obtained by linking the necessary libraries to the object files.

In contrast, the VPU technology operates on a host-based simulation incorporating the annotation of execution characteristics to reflect the behavior of the underlying hardware architecture (Sect. 7.3). This eliminates the need for cross-compilation, which is highly beneficial at early design stages where typically neither the compiler nor the final instruction set architecture is ready. Hence, the VPU directly enables the use of simulation techniques for early performance measurements, while simulation on the level of the instruction set needs to be delayed until the final architecture and the software tool chain is available. Additionally, the measurements performed by instruction set simulation are snapshots of the current development stage and might not reflect the final optimized software code. Thanks to the VPU, system architects can simply evaluate different design options by modifying the execution characteristic in minutes, while development and measurement of the software on the instruction set level might require anything from days to months.

To inspect the task modeling differences, a small example depicted in Listing 7.8 and 7.9 shall be consulted. The software code utilized for instruction set simulation is standard C-code. As the cross-compiler for the targeted architecture generates the binary, the driver implementations in line 10 and 16 of Listing 7.8 refer to load, and store operations to the corresponding memory addresses. These memory accesses are implicitly contained in the C source-code and generated by the compiler.

```
1
2
3
4  #define KEYBOARD_ADDR 0xE000;
5  #define DISPLAY_ADDR  0xF000;
6
7  void keyboard_driver(char *v) {
8
9
10     *v = *( (char*)(KEYBOARD_ADDR) );
11 }
12
13 void display_driver(char v) {
14
15
16     *( (char*)(DISPLAY_ADDR) ) = v;
17 }
18
19
20 void main() {
21   while (1) {
22     char value;
23     keyboard_driver(&value);
24     ...
25     // implement functionality
26     ...
27
28
29
30     display_driver(value);
31
32
33   }
34 }
```

Listing 7.8 Software intended for instruction set simulation

```
1  // includes standard task header
2  #include "MyTask.h"
3
4  #define KEYBOARD_ADDR 0xE000; #define DISPLAY_ADDR
   0xF000;
5
6  void MyTask::keyboard_driver(char *v) {
7    // reflect execution characteristic
8    consume(..);
9    read ( KEYBOARD_ADDR, v, size_of( char ) );
10 }
11
12 void MyTask::display_driver(char v) {
13   // reflect execution characteristic
14   consume(..);
15   write ( DISPLAY_ADDR, v, size_of( char ) );
16 }
17
18
19 void MyTask::task() {
20   while (1) {
21     char value;
22     keyboard_driver(&value);
23     ...
24     // implement functionality
25     ...
26     // reflect execution characteristic
27     consume(..);
28     ...
29     display_driver(value);
30     // reflect execution characteristic
31     consume(..);
32   }
33 }
```

Listing 7.9 Software intended for VPU-based simulation (C++/SystemC based)

When using the abstract simulation technology based on the VPU, these memory accesses need to be made explicit, since no cross-compilation step is performed. Within the illustrated example, the memory accesses are triggered by calling the explicit *read* and *write* memory access function of the VPU's HAL (line 10 and 16 in Listing 7.9). In order to be accurate, the VPU needs to be informed about the execution characteristic of each particular software piece. Therefore, developers annotate a particular characteristic with the *consume* function. This annotation follows one of the earlier introduced approaches of *statistical*, *source-level*, or *implementation*-based annotation (Sect. 7.3). For this small example, the corresponding tCEFSM of the given task can be captured consistently as illustrated in Fig. 7.13. Please note

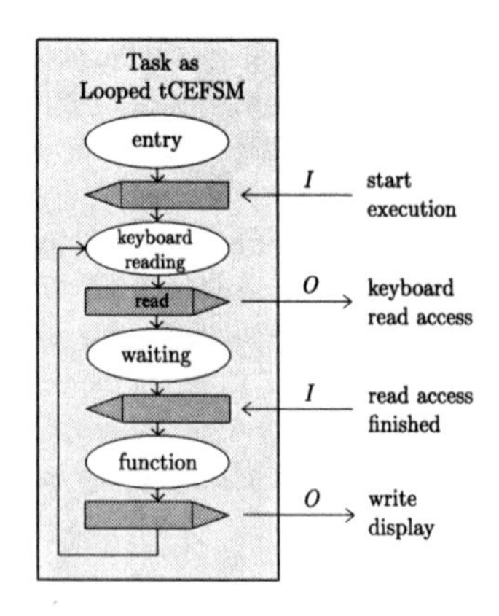

Fig. 7.13 Exemplary task model illustrated as tCEFSM

that, in general, it is possible to reflect each task by such tCEFSM. However, in real applications, the state number easily explodes and prohibits a simple visualization.

Apart from the annotation and the explicit memory accesses, the VPU simulation technology demands C/C++ or SystemC compliance. Hence, each individual task reflects a class that defines the actual procedure in the task function (line 20 in Listing 7.9).

The textual design entry basically does not restrict the software modeling any further and developers can easily program complex applications by making extensive use of functions from the operating system, including dynamic task management, middlewares, device drivers, and the hardware abstraction layer.

With respect to traditional software development, the textual design of an application introduces no restrictions. However, when applicable, modern design entries provide graphical visualization in order to simplify and speed-up the design process. These graphical design entries are especially well suited for applications based on process network and task graphs as discussed next.

7.5.2 Graphical Design Entry

The earlier discussion of applications pointed out that computational demanding wireless communication standards and multimedia applications can be consistently described as task graphs (Sect. 2.1). Hence, a graphical design entry similar to software component-based design [286–288] has been envisioned. Figure 7.14 illustrates the overall principle that will be discussed in the following. First the operational semantic is highlighted, followed by a discussion on the practical use.

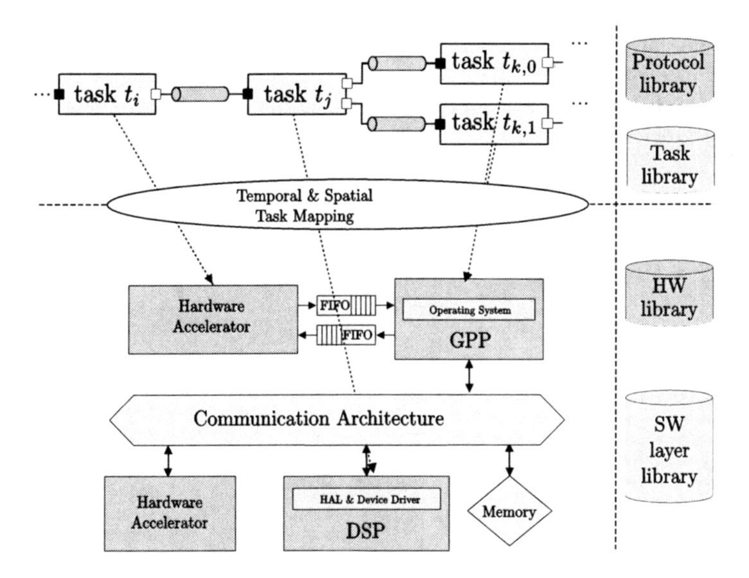

Fig. 7.14 Principle of the graphical design entry

The analytical implementation model defines a task graph as a 4-tuple $TG = (T, D, r, \delta)$ (see Definition 6.3). For a mathematical system evaluation, the previously introduced restrictions are highly beneficial to reduce analysis complexity. However, this is not required for simulation as dynamic effects can be handled naturally. Hence, an acyclic task graph and fixed data rates are *no* longer required; static scheduling techniques may be, but need not be, applied. This leads to a more generalized definition of the application similar to KPNs.

The generalized task graph definition removes some restrictions from the SDF task graph (Definition 6.3) which supports efficient mathematical processing.

Definition 7.2 (Application as Generalized Task Graph).

$$\begin{aligned} TG_{\text{generalized}} = & \ (T, D): \\ & T \text{ is the set of tasks } \{t_1, .., t_n\}, \\ & D \text{ is a set of ordered pairs } \{(t_i, t_j) : t_i, t_j \in T\}. \end{aligned}$$

The previous analytical implementation model defines a task only on the basis of its execution characteristic. Accordingly, the edges define data communication necessary to process the task and to emit data to another task. In simulation, this task description can and should be inspected on a more fine-grained level to achieve more accurate results and to allow investigation of detailed design issues. Therefore, a *looped tCEFSM* (Definition 7.1) defines a task within the graphical design entry. *Loop* behavior refers to tasks that execute over various states, but return to an explicit state z_i as depicted in Fig. 7.15. The implicit state kept inside the variables $\mathscr{U}$ can be of arbitrary value.

Each data edge $d_i \in D$ defines a data transfer between two tasks and is considered as a *channel* between two adjacent and connected tasks t_i and t_j (Fig. 7.14). With respect to a hardware abstraction layer (HAL) as introduced in Sect. 7.4.1 a task sends and receives data by utilizing the *put* and *get* functions. As illustrated, each put and get directly relates to a channel that corresponds to an edge for data exchange.

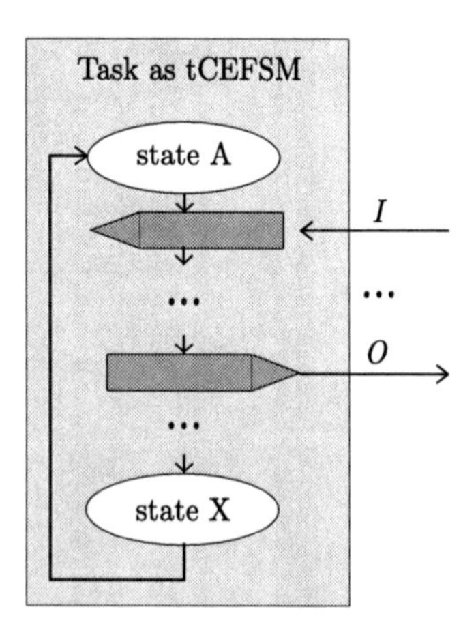

Fig. 7.15 Looped tCEFSM

This application description allows efficient software development in a graphical design manner based on the principle of task graphs. Assembling task graphs can be efficiently performed in graphical design tools, e.g., Platform Creator Tool (PCT) [173]. Besides the graphical representation by tasks and edges, the individual software development of each task is done on the basis of a textual design, e.g., the earlier discussed C/C++ based approach (Sect. 7.5.1). The clear interface separation (put and get) allows developers to focus on the behavior, while data communication is handled by the channel and its respective protocol implementation.

Up to this point, only the application part has been considered for the graphical design entry. However, for performance evaluation of the complete system, the hardware platform as well as the temporal and spatial task mapping must be incorporated. From an abstract point of view, each hardware platform, in particular any MPSoC platform, consists of processing elements, communication architectures, and memories. Accordingly, a 3-tupel $HW = (PE, CA, Mem)$ (see Definition 6.5) defines any hardware architecture [251]. The class of processing elements contains all kinds of components such as GPPs, DSPs, ASIPs, but also hardwired accelerators. The VPU simulation model is able to emulate all of these processing elements, as demonstrated previously (Fig. 7.2). The combined usage of various SystemC components, including VPUs, allows for a complete system simulation.

Having specified a hardware platform candidate and having assembled a simulation model, the temporal and spatial task mapping can be performed. The graphical representation of the temporal and spatial task mapping denotes (1) the task-to-processor assignment (spatial mapping) and (2) the execution order of tasks on one processing element (temporal mapping). The spatial task mapping (STM) as given in Definition 6.7 represents a simple drag and drop method of tasks to hardware components within a graphical environment (Fig. 7.14). With respect to temporal mapping, the static schedules (Definition 6.8) can be easily visualized, which is not feasible for dynamic schedules defined by an underlying algorithm, e.g., round-robin or priority based.

With the mapping of a task graph onto a particular hardware platform, the channels utilized for data exchange need to be instrumented by protocols. This protocol instantiation requires an existing physical link between the corresponding processing elements.

Definition 7.3 (Application-to-Architecture Mapping in a Graphical Environment). The application-to-architecture mapping consists of the following steps.

1. **Task Development Phase.** Each task is developed based on the textual design entry (Definition 7.5.1) with the optional use of software features, e.g., the operating system. Data exchange is restricted to the use of dedicated channels. Hence, task modeling is kept independent from the mapping and from the protocol for data exchange by the use of clearly specified interfaces (put and get).
2. **Spatial Task Mapping Phase.** As specified by Definition 6.7 the application tasks are mapped onto the processing elements of the hardware platform (Fig. 7.14), while obeying the restrictions given in Definition 6.6.

3. **Temporal Task Mapping Phase.** For each processing element a scheduling (temporal mapping) needs to be defined that follows either a static (Definition 6.8) or dynamic scheduling scheme.
4. **Protocols Implementation Phase.** The tasks exchange data by the use of explicit interfaces (put and get) and dedicated channels, represented as edges within the task graph (Fig. 7.14). However, these channels need to be instrumented by protocols, e.g., those specified in the hardware abstraction layer (Sect. 7.4.1).

$$\begin{aligned}
\exists protocol\ \forall d_k \in D = \{(t_i,t_j) : t_i,t_j \in T(DFG)\} :&\\
&\text{with } protocol = (data, location, features),\\
&\text{with } data\text{, e.g., variable, fifo,}\\
&\text{with } location\text{, e.g., shared memory, hardware fifo,}\\
&\text{with } features\text{, e.g., blocking, nonblocking.}
\end{aligned}$$

Following these phases, system architects obtain a complete VPU-based simulation model which serves for evaluation of decisions at early design stages. The VPU technology and the graphical design entry have been successfully transferred into a commercially available tool [171].

Acceptance and usability are mandatory for design space exploration tools. Hence, the level of abstract simulation connects smoothly to the lower instruction set simulation-based model and is sketched in the following section.

7.6 Refinement to Instruction Set Simulation

The overall design methodology proposes a continuous flow from the analytical model to the final implementation. Within the refinement flow, the abstract simulation model is located between the analytical and the ISS-based implementation model. As introduced earlier, the Time Retrieval Engine closes the gap from the abstract simulation model to the above located analytical model. Apart from coupling these two abstraction layers, the discussion now turns to the link to lower abstraction levels like instruction set simulation.

The results of the abstract simulation-based design space exploration are identified implementation candidates which meet the necessary application constraints. However, the ultimate goal of any design process is the final implementation, consisting of software and hardware. As the abstract simulation model operates on the principle of abstracting the underlying hardware, system architects have to lower the abstraction level for more fine-grained system evaluations. Naturally, this path leads to the final implementation. Refinement in itself impacts the two major components given by software and hardware. The subsequent discussion starts with the refinement of the hardware simulation model. Later, it introduces the porting of the software from the abstract VPU simulation model to the instruction set simulator.

7.6.1 Hardware Simulation Model Refinement

In general the hardware architecture of any MPSoC follows the design principle of component-based design. When using this design method, system architects assemble various IP components to form the complete hardware architecture. A rough classification separates these components into *processing elements*, *communication architectures*, and *memories* [251,288]. Development of simulation models at Electronic System Level (ESL) also follows this principle. Here, the simulation model of the complete platform is assembled by putting the simulation model of all components together. The resulting system simulation model is usually referred to as *virtual platform*.

As discussed earlier, development of the component models typically uses the SystemC programming language. The use of the Transaction Level Modeling standard 2.0 [153] separates the component model from the external system by defining a clear interface structure. This separation allows independent development of a simulation model for each component. Because of their reactive (slave) behavior, development of memory simulation models is done as for standard SystemC components with a particular delay period. Therefore, in recent years research has focused on the communication architectures and the processing elements.

Besides specific communication architecture models, e.g., AMBA AHB or AXI [34], generic models such as the ones contained in the Architects View Framework (AVF) [282] allow quick evaluation and exploration of different design options. As the VPU is based on SystemC and is compliant to the TLM 2.0 standard, the same design options exist for virtual platforms. Hence, this only leaves the processing elements emulated by VPUs subject to refinement.

The VPU emulates the behavior of an underlying processing element within the system simulation. This includes effects of software execution and the emulation of arbitrary communication including the interfaces and hardware ports. With each component being capsuled by the communication interfaces, it is a black box for the other components and data exchange only occurs at these interfaces. Hence, refinement of the VPU to an instruction set simulator is reduced to a simple replacement of the configured VPU with the dedicated instruction set simulator as depicted in Fig. 7.16. Of course, this assumes the existence of the simulator and its compliance to the TLM-2 standard.

The replacement of the VPU has a stronger impact on the software development and implementation than on the hardware simulation model. This impact is discussed next.

7.6.2 Software Refinement

Basis of today's software development is the use of abstraction layers. Reasons for using them are manifold. On the one hand the abstraction layers shield software developers from lower level details by hiding fine-grained implementation details.

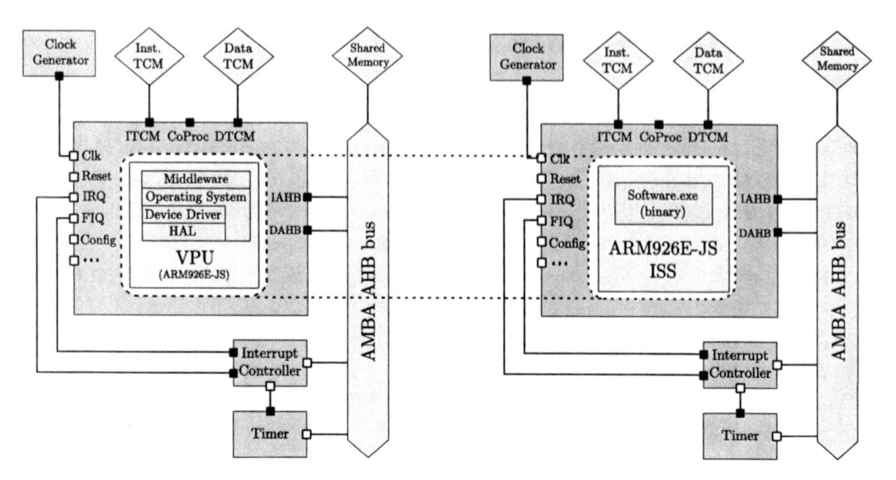

Fig. 7.16 Refinement example – from VPU to ISS

Further, porting applications from one platform to another one is simplified as only hardware specific software layers require modification. To enable the same software development for the VPU as for the real hardware, the VPU model comes with the discussed set of software layers that are:

- Middleware libraries
- An operating system
- Several device drivers
- A hardware abstraction layer

When modeling a particular application composed of different tasks, the only difference to C/C++ programming is the requirement for annotating the execution characteristic. Additionally, explicit memory accesses have to be used because no cross-compilation and no instruction set simulation is performed. Therefore, software refinement is equivalent to porting an application from one platform to another with reduced effort, as only the following steps need to be applied.

1. Replacement or platform-specific implementation of all explicit memory access calls.
2. Replacement of the generic OS by a specific one.
3. Removal of all annotations of the execution characteristic since instruction set simulation executes the executable instruction-by-instruction.

The last step characterizes the removal of all *consume* calls as well as software that relates to the computation of these annotations and can be automated. The other refinement steps cause more development effort. When modeling software on top of the VPU, memory accesses have been made explicit by calling the *read* and *write* functions of the hardware abstraction layer (Sect. 7.4.1). With the replacement of the VPU by an instruction set simulator, this has to be reversed as the compiler inherently emits load and store instructions. Hence, two options exist in replacing

Table 7.4 Replacement or implementation of explicit memory access functions

Explicit memory access in abstract simulation model	Corresponding C implementation for Instruction set simulation

write(unsigned int address, int data, int size)	unsigned int *address=...;
	for (int i = 0; i<size; i++) {
	    *(address+i) = data[i];
	}

read(unsigned int address, int &data, int size)	unsigned int *address=...;
	for (int i = 0; i<size; i++) {
	    data[i] = *(address+i);
	}

the explicit memory accesses. First, the read and write functions are ported to the specific hardware platform. Since this approach causes overhead, the better option is to replace the function calls by the corresponding software code directly within the source code. Table 7.4 illustrates an exemplary replacement of a read and write function with integer data and address types.

The earlier steps are sufficient for the low-level software given by the *hardware abstraction layer* and the *device drivers*. However, the operating system layer requires additional effort during refinement.

Within the discussion of application and hardware platforms, the fragmented domain of available operating systems has been sketched. As a consequence the applied generic OS running on the VPU has been defined on the basis of process management as well as interprocess communication and synchronization, rather than replicating a single operating system. Hence, the software code must be ported to the specific OS that later executes on the targeted platform. To simplify this porting, correspondence tables determine the refinement path between the generic OS and a specific one.

Table 7.5 reflects the mapping of some important operating system function calls, which exemplifies the mapping principle. It should be noted that, the functionality

Table 7.5 Specific OS API refinement of important OS functions

Generic OS API	RTEMS API	POSIX API
OsTaskId	rtems_task_ident	getpid
OsTaskSuspend	rtems_task_suspend	pthread_kill(..., SIGSTOP)
OsTaskResume	rtems_task_resume	pthread_kill(..., SIGCONT)
OsTaskYield	rtems_task_suspend	sched_yield
OsTaskYieldTo	rtems_task_resume rtems_task_suspend	pthread_kill(..., SIGCONT) pthread_kill(..., SIGSTOP)
OsTaskCreate	rtems_task_create	pthread_create
OsTaskDelete	rtems_task_delete	pthread_exit
OsTaskStart	rtems_task_start	–
OsWakeAfter	rtems_task_wake_after	sleep
OsShutdown	rtems_shutdown_executive	exit
...	...	...

```
1  void task() {
2    unsigned int id=0, p=0, old_p=0;
3    unsigned int size=4, comm_id=1;
4
5    OsTaskId(&id);
6
7
8    OsTaskGetPriority(id, &p);
9    p++;
10   OsTaskSetPriority(id, p, &old_p);
11
12
13   while (data < 50) {
14     put(comm_id, &data, size);
15     OsWakeAfter(150);
16     data+;
17   }
18   OsTaskDestroy(id);
19 }
```

Listing 7.10 Generic OS API

```
1  void task() {
2    unsigned int id=0, p=0, old_p=0;
3    unsigned int size=4, comm_id=1;
4
5    rtems_task_ident(RTEMS_SELF,
6              RTEMS_SEARCH_ALL_NODES, &id);
7
8    rtems_set_priority(id, p, &old_p);
9    p++;
10   rtems_set_priority(id, RTEMS_CURRENT_PRIORITY,
11             &prio);
12
13   while (data < 50) {
14     put(comm_id, &data, size);
15     rtems_task_wake_after(150);
16     data++;
17   }
18   rtems_task_delete(id);
19 }
```

Listing 7.11 RTEMS API

Fig. 7.17 Operating system specific refinement example

of the generic OS API call has to be reflected by the functionality of the specific API. Thus, generic API calls might be replaced with *no*, *one*, or *several* specific operating system API calls. An example where the API call is simply removed, is the generic *OsTaskStart* function when translated to the POSIX standard. Here, the tasks are inherently set to running state, hence starting the task is not necessary. On the other hand, the *OsTaskYieldTo* function is mapped to two underlying specific functions that together implement the same functionality.

This replacement of the generic OS calls is illustrated by an example in Fig. 7.17. First the task identifies its current priority (line 1-6) and increments the priority by one (line 8 + 9). Then the task sends 50 times a data sample (*put* in line 10) to the next task. After each sent the task sleeps for 150 time-units. Finally, it destroys itself (line 13).

On the basis of the presented technique, all generic OS API calls are translated to their specific representatives allowing software developers to model their complete application in a quick and simple manner. Additionally, portability and reuse of the developed application are ensured by the generic OS API.

Finally, the middleware that operates on top of all other software layers should not be affected or, at most, minor modifications due to explicit memory accesses may be required as previously discussed. The same applies to the application task developed on top of all abstraction layers.

Summarizing, the introduced software refinement from an abstract model to an instruction set simulation model is rather straightforward and requires only minor changes. Therefore, it was possible to develop an automatic refinement flow for implementation candidates that follow the graphical design entry (Sect. 7.5.2).

7.6.3 Automatic Refinement Flow for the Graphical Design Entry

The information contained in the graphical design entry allows for an automatic refinement flow. Figure 7.18 illustrates this flow centered around the Platform

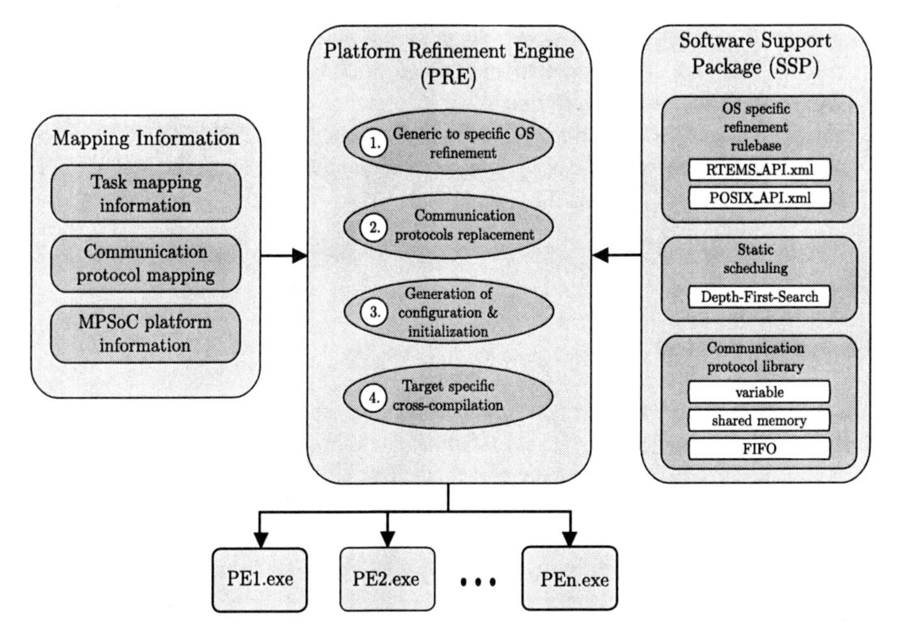

Fig. 7.18 Platform refinement engine (PRE)

Refinement Engine (PRE) that is described later. The complete refinement can be divided into four different phases:

1. *Generic to specific operating system refinement.*
   In case a specific operating system is chosen, the PRE iterates over all tasks and replaces the generic OS API calls by the specific ones. The replacement rules for a specific OS are stored in a library, named the *Software Support Package (SSP)*. Currently only a small set of libraries exist, e.g., one for the RTEMS OS, but the modular structure allows easy integration of new libraries in the future.
   Besides the use of dynamic scheduling based on an operating system, a static scheduler can be selected. However, for an automatic computation of the schedule, a restriction is imposed that applications need to be modeled as SDF graphs. In this event, the generic OS API calls are removed as the tasks will be called by the scheduler in sequential order.
2. *Communication Protocol Replacement.*
   Based on the defined communication protocols, the generic put and get interface calls are replaced. Again the replacement rules are kept within the SSP's communication protocol library.
3. *Generation of the configuration and initialization.*
   In the third phase, the PRE generates the configuration and the initialization task. These initialization tasks are either the scheduler task or the operating systems initialization task which comprises, e.g., the maximum number of tasks, message queues, etc. Most of these parameters can be extracted from the two previous

phases. Nevertheless, developers can tune the configuration to ensure optimal performance, e.g., the computed scheduling might be hand-tuned if necessary.

4. *Target-specific cross-compilation.*
   Finally, the generated source code is cross-compiled. The resulting software executable can then be used for system simulation and/or later on the real MPSoC hardware.

The following section highlights the operating system specific refinement.

OS Specific Refinement

Today a wide variety of different OSs and RTOSs are available [275]. Therefore, software developers have to identify the most suitable operating system which may be a general-purpose or a more specific Real-Time Operating System (RTOS).

In the presented approach, the key idea is to raise software development to a higher abstraction level. Software development is based on a generic OS API. Later, this generic OS API will be replaced according to the specific targeted OS running on the processor core. Currently, refinement to the RTEMS and POSIX API is supported. Extending the design flow to other OSs is straightforward. For each OS a specific rule base in XML [289] format only has to be developed *once*.

As highlighted earlier, the function replacement follows a predefined correspondence table. Therefore, the automatic refinement replaces all function calls of the generic OS API by the corresponding specific ones. This allows easy and quick porting of any VPU and generic OS software to a specific operating system. However, it should be noted that special features of the targeted operating systems might not be optimally exploited. Therefore, the proposed flow enables developers to hand-optimize the generated code later on.

Communication Protocols

The communication channels between tasks are instrumented by communication protocols being part of the hardware abstraction layer. During software refinement, all utilized software layers, including the hardware abstraction layer, are ported to the targeted hardware platform. These implementations are stored in the software support package and are ready for use. The selection of the communication protocol defines which protocol of the HAL is instantiated later on. In order to minimize the implementation overhead, the PRE inlines these protocols into the source code.

Configuration and Initialization

The last two phases of the PRE are the generation of the configuration and initialization, followed by the final cross-compilation.

When using a static scheduler, the configuration step consists of computing the static schedule. In case of an operating system is being incorporated, the configuration step comprises the setting of parameters, e.g., the maximum number of tasks or message queues. In principle this is rather operating system specific and typically requires fine tuning by the software developer. Nevertheless, the PRE generates a template for the software developer on the basis of the input of the first two phases, i.e., the number of tasks can be computed by the mapped tasks and their defined instances.

Typically, initialization of an operating system is performed during bootup by calling an initialization task. For the supported operating-system APIs, this task is automatically generated and has the following structure. First, the operating system itself is configured according to the configuration parameters, e.g., the scheduler policy such as round-robin scheduling. Second, all mapped tasks are created and started within the operating system. Afterward all communication protocol specific parts are created, e.g., the message queues.

Finally, all refined tasks and the initialization task are cross-compiled and linked together with the specific operating system libraries to obtain the software executable for the underlying processor cores.

The software executable can then be used for simulation in the refined hardware simulation model. This directly allows evaluation and exploration at the level of instruction set simulation.

7.7 Summary of the Abstract Simulation Model

Throughout this chapter the abstract simulation model including the Virtual Processing Unit has been presented. A brief summary is given later.

- The *Virtual Processing Unit* with the annotation principle of execution characteristics forms the centerpiece of the proposed simulation technique. The generic nature of the VPU allows emulating all types of processing elements, from pure hardware centric accelerators, via specialized processor cores to general purpose cores. As the precision of the annotation technique heavily depends on the system architect's knowledge of the application and hardware architecture, different annotation techniques are supported.

  - Statistically based timing annotation.
  - Source-level-based timing annotation.
  - Implementation-based timing annotation.

  These timing annotation techniques ensure that system architects can always incorporate the best knowledge available to achieve the most reliable results at particularly early design stages. Additionally, the path from coarse- to fine-grained timing annotation is given to guide system architects in their design process.

- Software development generally follows a pragmatic design approach, making extensive use of software libraries and abstractions. Hence, the framework provides a set of different *software layers* in addition to the pure abstract VPU simulation model. These subsume the most common ones like hardware abstraction layers, arbitrary device drivers, the operating system, and other middleware libraries.
- Together the VPU and the different software layers provide an efficient environment to model complex applications and hardware platforms for simulation. As software development is becoming a central component in today's and future MPSoC designs, two different design entries are supported. The *textual design entry* is closely related to traditional software development, while a *graphical design entry* allows efficient software development and modeling by task graphs and process networks.
- The abstract simulation model integrates smoothly into commercially available tool flows by a well-defined link to lower levels of abstractions. In addition, a (semi-)automatic refinement flow has been introduced to efficiently bridge the gap between abstract and instruction set based simulation.

Chapter 8
Case Study

A case study proving feasibility is always key when proposing new design tools and methodologies. It can also clarify the advantages but also limitations of the proposed design methodology.

The chapter is organized in two parts. The first part presents a case study at the level of individual tasks in order to identify the modeling accuracy of the proposed annotation methods. The discussion includes various hardware architecture considerations to encompass a complete design picture. From the application point of view, the focus is on typical mathematical algorithms rather than on a single specific application.

After discussion of the achievable accuracy of the annotation methods, the second part of the chapter highlights the practical use of the complete framework. To illustrate the work flow, the case study targets the domain of wireless communication, especially SDRs. The major contribution is the investigation of a complete design process starting at a high abstraction level and continuing along the proposed design method toward a final implementation.

8.1 Task Level Annotation

Section 7.3 has introduced the various annotation techniques in order to characterize a VPU according to any given architectural processing element.

- Statistical annotation.
- Source-level annotation subdivided into:
  - Coarse-grained (e.g., on basis of functions).
  - Fine-grained (e.g., on basis of basic-blocks or source code lines).
- Implementation-level annotation subdivided into:
  - Profiling-based (e.g., µProfiler).
  - Trace-based.

T. Kempf et al., *Multiprocessor Systems on Chip: Design Space Exploration*,
DOI 10.1007/978-1-4419-8153-0_8,

An optimal annotation model exists for each architectural processing element, ranging from general purpose processors to hardwired accelerators. Specific hardwired components, such as ones from Synopsys DesignWare [157] or other IP vendors, typically have a fixed and precisely characterized behavior based on a reactive execution scheme. Past research has demonstrated that such components can be efficiently modeled based on the principle of VAM [282]. Similarly, a single VPU can easily capture such scenarios by executing one application task with a particular execution characteristic.

In general, the VPU targets more complex use cases where multiple applications are executed on a shared processing resource, e.g., multiple applications running under the control of an operating system. These require much more advanced annotation techniques to emulate the execution behavior correctly. Due to the variety of possible design options, the proposed framework provides different annotation methods so that the best technique can be selected for each addressed use case. Of course, this introduces the challenge to select the right method for the targeted hardware architecture and software implementation. To identify the best-suited one for each use case, a thorough analysis presents common scenarios and inspects the precision of the annotated VPU simulation model in comparison with cycle accurate models.

8.1.1 Task Level Analysis Scenario

Focussing jointly on complex scenarios combining software and hardware issues, the following case study investigates the key issues:

1. Hardware architectures, in this case only programmable processor cores.
2. Algorithmic implementations.
3. Implementation options, whereby only software implementations (C/C++ or Assembler) are considered.

Though addressing complex scenarios, only programmable architectures are considered. In order to capture these, several General Purpose Processors (GPPs) and Digital Signal Processors (DSPs) have been selected to identify suitable annotation techniques. From the domain of GPPs, the ARM720T and ARM926EJ-S [174], have been selected and from the DSP domain, TI's C55x [290] and C64x [291].

The chosen applications cover a large variety of fundamental mathematical algorithms which have been extracted from TI's DSP libraries [292].

- Vector operations, e.g., vector addition and product.
- Matrix operations, e.g., transposition and multiplication.
- Filter operations, e.g., FIR and adaptive LMS filtering.
- Correlation operations, e.g., autocorrelation.
- FFT operations, e.g., FFT and IFFT.

When considering MPSoC platforms in the domain of wireless communications, the implemented applications impose tight constraints on high performance and energy efficiency, especially for mobile and battery powered devices. Accordingly, this demands fully exploiting the available hardware features. As a consequence developers *must* optimize given applications for the underlying hardware architecture to minimize the induced overhead. This results in a diverse environment for software development. Targeting general purpose processors, high-level programming languages like C are preferred; whereas DSPs require more hardware centric software design. There are many reasons for this, mostly relating to compiler issues [293, 294]. The case study within [60] identifies the resulting performance gap between hand-optimized Assembly and general purpose C-code for the C64x DSP to be larger than one order of magnitude. For the smaller C55x DSP, this difference is determined to be a factor of approximately eight. However, the difference depends on the application, the coding style and must therefore be considered on an individual basis. In addition, the development effort spent on Assembly programming should not be forgotten and hence always defines a trade-off decision between performance and costs.

The ultimate goal of any design-space exploration technique is to identify the performance of the final implementation in order to give design hints at the current development stage. Accordingly, the annotation techniques must mimic the execution characteristic of the *final implementation* rather than an intermediate snapshot, resulting for example from the execution of a reference implementation on a DSP with far worse performance than the finally optimized software. Therefore, the discussed task-level analysis aims at realistic final implementation candidates to demonstrate the capabilities of the proposed annotation techniques. The investigated implementation options are illustrated in Table 8.1.

Figure 8.1 depicts the fundamental principle to evaluate the accuracy and precision of any annotation technique. The reference is an implementation of each application, optimized for the utilized hardware architecture and executed on a cycle accurate instruction set simulator. During execution each *memory access* and the *execution time* in cycles are measured.

Each reference implementation is compared to the trace collected with the VPU simulation model including the annotation technique subject to evaluation. This comparison is done with respect to *memory accesses* and required *clock cycles*. The obtained results reflect the accuracy and errors of the annotation technique for the particular implementation scenario.

As discussed earlier various annotation techniques exist, of which two operate on the *implementation level*. The *trace*-based annotation technique replays the traced

Table 8.1 Considered task level analysis scenario implementation options

Processor type	Processor core	High-level software	Low-level software
GPP	ARM720T	X	–
	ARM926EJ-S	X	–
DSP	TI C55x	–	X
	TI C64x	–	X

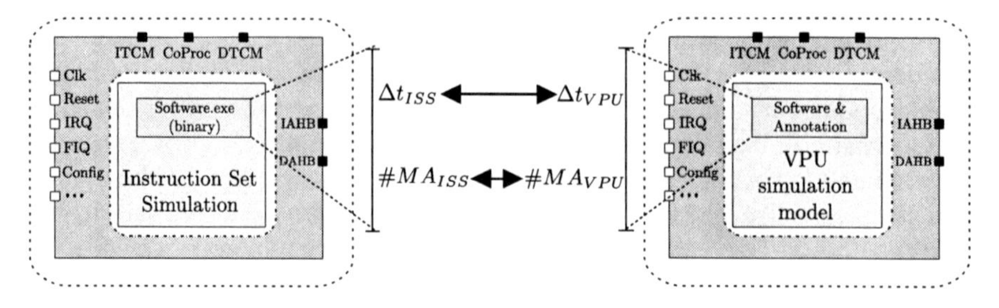

Fig. 8.1 Evaluation principle of annotation techniques (ARM926EJ-S example)

instruction set simulation using the VPU technology. This definitely replicates the execution characteristic in terms of execution delay and memory accesses exactly. Therefore, *no* error exists as the same trace is taken as reference.

The other implementation-based technique is *profiling* based. The particular implementation discussed in Sect. 7.3.3 utilizes the μProfiler [255] that operates on the LANCE compiler tool chain [295]. Other techniques such as TotalProf [296] and other annotation techniques [297] utilize compilers like LLVM [271]. The promising results at least for C-based software and RISC-based architectures have attracted quite a lot of research activities in this area. These techniques require – similar to instruction set simulation – a software implementation that can be measured, but may not be available during early design space exploration. Profiling-based techniques should only be considered when the underlying processor core is not known, the instruction set simulator, or the compiler tool chain is not yet finished. Otherwise, trace-based annotations should be preferred because they achieve higher accuracy.

Because of these issues, the subsequent investigation focuses on the statistical and source-level annotation models and their evaluation.

8.1.2 Task Level Analysis Results

So far the discussion has highlighted the principle for the evaluation of annotation techniques in general. To identify the achievable accuracy and the incurred error of annotation methods, the introduced case study inspects the modeling accuracy of annotation models for the example implementations introduced in Table 8.1.

Annotation Results

Following the proposed design process, system architects need to start the final implementation at a particular point in time. Typically, this starts with a pure algorithmic implementation in C programming language or MATLAB [298]. This reference implementation can be utilized to verify functional correctness with the

VPU simulation model together with a realistic evaluation of the performance characteristics. At later design stages, the increased implementation knowledge enables system architects to incorporate more precise characterizations.

To demonstrate this process, the case study investigates coarse- and fine-grained annotation techniques in the following. Targeting general purpose processor cores that are programmed in a high-level programming language, the provided reference implementation can help to identify common software constructs, e.g., loops and if-else cases. The functionally correct execution ensures that only the right annotations are added at run-time. This makes fine-grained source-level annotation methods the optimal choice for such implementations (Sect. 7.3.2). These annotations can be applied at the granularity of functions, basic blocks, or even source code lines to keep track of the execution behavior.

In contrast to this, processor cores with hardware centric software implementations tend to require different techniques. This does not necessarily mean that fine-grained source-level annotations are false. However, these annotations operate on a reference software implementation developed in a high-level language which most likely does not reflect the behavior of the targeted final implementation. Especially, when targeting highly specialized architectures, the final implementation might achieve much higher performance than the implementation based on the reference [60]. Therefore, coarse-grained statistical and source-level annotations are preferable for such architectures. In addition, at the time of instrumentation, system architects should recollect the fundamental mathematical operations within the application and underlying algorithm. For example, when targeting a vector dot product on TI's C64x DSP, the mathematical operation defines a multiplication per vector elements. On the assumption of 16-bit per vector element, the processor core is able to load four elements of each vector at a time (2x LDDW instruction) and four output vector elements can be computed per cycle (2x DOT2P instructions). By applying a software optimization, pipeline stalls can be prevented and the resulting execution cycles compute to $NX/4 + overhead$ with NX the number of input vector elements. Such considerations can exactly replicate the final implementation and hand-optimized Assembly code, also done in TI's DSP library [292].

Based on these investigations, the two design options have been emulated differently; on the one hand the GPP architectures are targeted with a VPU simulation model and fine-grained annotations, and on the other hand a coarse-grained annotation has been selected for the DSPs.[1] Figures 8.2–8.4 illustrate the results of the measurements. As visible, the occurring error varies slightly for the different algorithms.

In a simplified investigation, the overall execution characteristic is determined by two major aspects. First and most important, the execution cycles spent in the computation of the processor core are determined by the core cycles. Highlighted

[1] Please note that for the coarse-grained annotations, the objective is the hand-optimized assembly code that has been determined to execute more than one order of magnitude faster than the general purpose C code [60]. Clearly this demands the optimized software as a reference implementation, otherwise the estimations would be far off reality (~8–10 times).

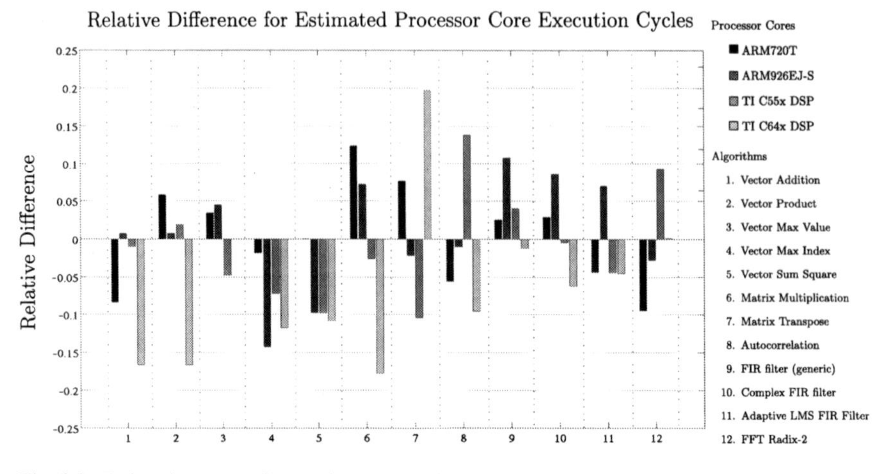

Fig. 8.2 Estimation error of execution core cycles

in Fig. 8.2, the deviation for the fine-grained instrumentation of the GPP execution characteristic has been measured as a maximum error of 14.3% and as an average error of 5.6%. For the DSP architectures, a maximum error of 19.7% and an average one of 7.7% has been measured when using the suggested coarse-grained instrumentation technique.

The second aspect, the communication aspect, is investigated separately for the program and data memory accesses. The error is measured as the difference between traced memory accesses when using the instruction set simulator and the VPU-based simulation. The results for the investigated algorithms are illustrated in Figs. 8.3 and 8.4. While the first Fig. 8.3 demonstrates the deviation for program memory

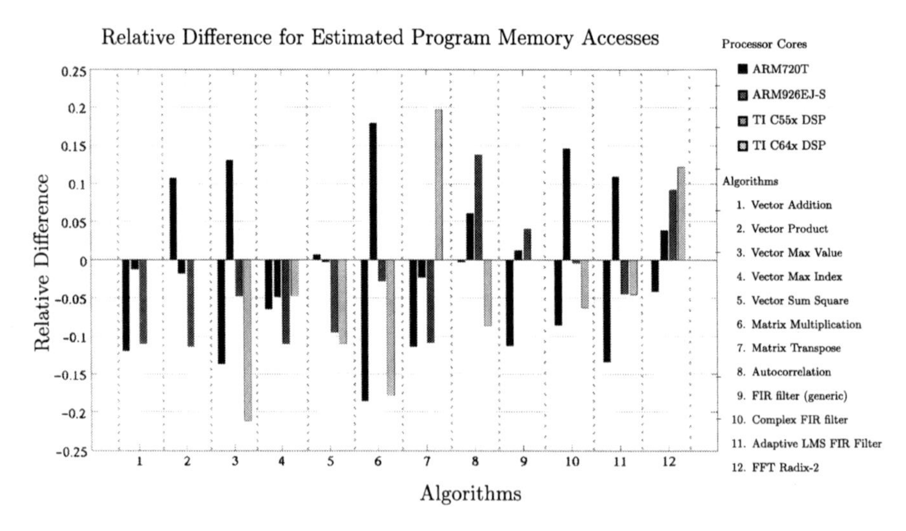

Fig. 8.3 Estimation error and accuracy of program memory accesses

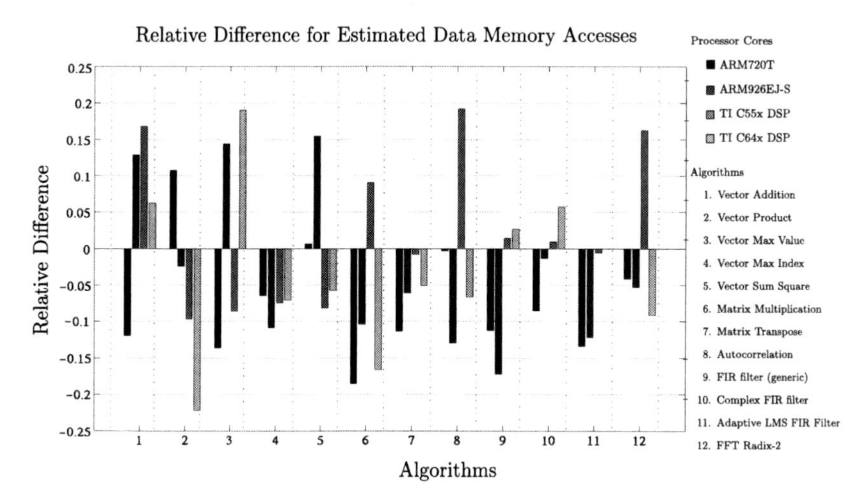

Fig. 8.4 Estimation error and accuracy of data memory accesses

accesses, the second Fig. 8.4 highlights the incurred error for data memory accesses. The encountered maximum error is less than 22% which is sufficient for early design phases.

Summarizing the results of the inspected annotation results, the determined errors are totally acceptable for early design space exploration where the number, type, and interconnection of processing elements is the subject of evaluation. However, system architects need to carefully select the right method when addressing a particular implementation. For example, when addressing DSPs and low-level software implementations, coarse-grained instrumentation techniques are beneficial, while fine-grained annotations might show misleading results.

Summary of Task Level Analysis

The presented analysis has shown that carefully applied annotation techniques can emulate the execution characteristic of arbitrary tasks precisely enough for early design stages. In addition, the exemplary analysis highlighted how increasing implementation knowledge can enhance modeling of the execution behavior in terms of precision.

However, the demand for multiple annotation techniques has been demonstrated in order to reflect the final implementation behavior and not a preliminary result. The discussed techniques show precise modeling for the purpose of early design space exploration when selected for the right final implementation. This makes selection of the correct annotation principle for the targeted implementation critical, as already shown and answered earlier. For clarification Table 8.2 illustrates a compatibility matrix from the gathered experiences that helps system architects to select the optimal annotation model for the intended implementation.

Table 8.2 Compatibility matrix for annotation techniques of the execution characteristic

Targeted final implementation	Statistical model	Source-level models		Implementation models	
		Coarse-grained	Fine-grained	µprofiler	Trace-based
GPP and C-code	+	+	++	+	++
DSP and C-code	+	++	+	∘	++
DSP and Assembly	+	+	++	∘	++
ASIP and assembly	+	++	+	∘	++
HW accelerator	+	++	+	−	++

++ = recommended, + = applicable, ∘ = applicable, but not recommended, − = not applicable

8.2 System Level Case Study

So far, accuracy and achievable precision of arbitrary task level annotations has been discussed. In the following of this section the capabilities and the workflow of the design space exploration framework are exemplified with a case study from the domain of wireless communication. The case study derives from a realistic project goal of an SDR development.

The main objective defines the design of a receiver hardware platform optimized for two wireless communication standards. The platform shall be assembled mainly out of standard IP components to minimize development effort and costs. Besides the tight hard real-time constraints of the communication standards that must be kept at run-time, energy efficiency should be maximized.

To display a realistic scenario, the two inspected wireless communication standards represent a low and high data rate standard. The first communication standard is derived from the physical layer of the *MIL-STD-188-110B* standard that implements a robust low data rate standard [299]. The second standard to be implemented is composed of standard signal processing blocks and will be denoted as a *Representative Communication Algorithm* in the following.

After the introduction of the fundamental algorithms and wireless communication standards, the hardware IP components available for the development of the platform are introduced. Finally, the envisioned design flow based on the exploration framework is exemplified. At the start of the design, the analytical model is utilized to identify potential candidates and the abstract simulation technique is jointly utilized to validate the obtained results.

8.2.1 Wireless Communication Standards

MIL-STD-188-110B

Initially developed for military applications, the MIL-STD-188-110B defines a low data rate, but a highly robust communication standard. It provides a set of different data rates ranging from 75 to 9,600 bit per second. Accordingly, the standard defines

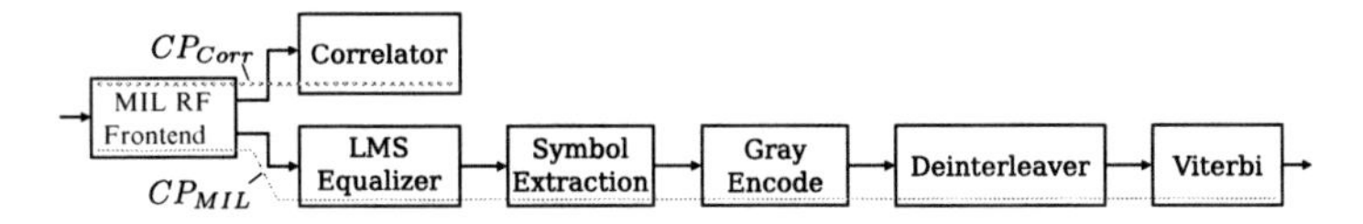

Fig. 8.5 The MIL-STD-188-110B algorithm

multiple modes, each mode given by a set of parameters, e.g., the data rate, code rate, and interleaver length. For simplification during this case study, focus is set on only one mode of the standard. The inspected mode is a slightly modified transmission scheme that operates on 4,800 symbols per second instead of the 2,400 symbols per second defined by the original standard. Transmission utilizes an 8-symbol phase-shift keying in the original and the modified implementation.

Data transmission is performed on the basis of frames, which are assembled in four stages. Each frame starts with a preamble for synchronization, followed by user data, a training sequence, and ends with a sequence indicating the end of message. Focussing on the receiver part, the task graph for the implementation is highlighted in Fig. 8.5. The task processing chain has been separated into two phases, namely the correlation and decoding. The initial state is defined by the correlation in which the receiver algorithm processes the received data to detect the start of a transmission. After the preamble sequence has been detected, the algorithm switches from correlation to decoding stage, finally emitting the received data.

Both phases and the involved tasks operate separately from each other, which allows both illustrated critical paths within the task graph to be evaluated individually. The highlighted critical paths under evaluation are sensitive to throughput constraints, while latency constraints are rather uncritical. The throughput constraint is typically given as samples per second. However, to minimize computing effort, the case study operates on the inverse, the sample processing time. For the evaluated mode this sample processing time computes to

$$\frac{t}{\text{sample}} < \frac{1\,\text{s}}{4{,}800} = 208.3\,\mu\text{s}.$$

This sample processing time defines the fundamental scheduling period in which the same task instance needs to be executed twice. Otherwise, resulting buffer overflows will inevitably discard received data.

Within the following case study the tasks are shortened as Correlation `(Corr)`, Equalizer `(EQ)`, SymbEx `(SE)`, GrayEncoder `(GE)`, Deinterleaver `(MDI)`, and FECDecoder `(FEC)`.

Representative Communication Algorithm

The Representative Communication Algorithm (RCA) is not standardized, but reflects the common features of communication standards utilized today. Figure 8.6 highlights the defined receiver of the physical layer processing. It defines a high

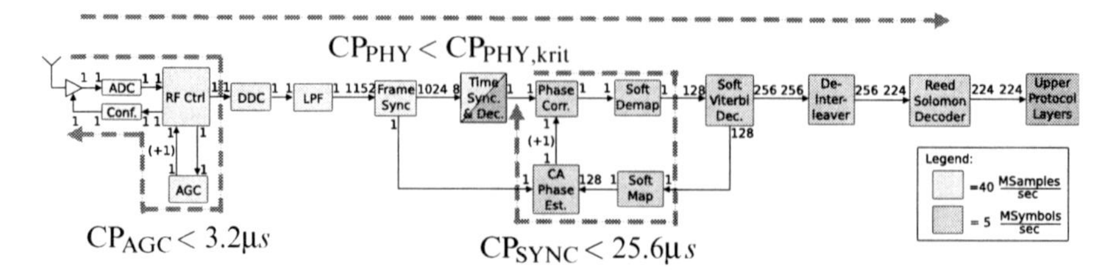

Fig. 8.6 Representative communication algorithm

data rate communication with feedback loops and typical tasks for signal processing. In contrast to the previously discussed MIL-STD-188-110B standard, the RCA includes the two phases together and executes on each received sample.

The RCA checks each received sample for the beginning of a data frame. Whenever a frame has been detected, the data is processed by the following algorithmic tasks. After correction of the timing and phase offset, symbol demapping and decoding of the convolutional code are applied. The resulting soft information allows improvement of the estimation of the phase offset to finally reduce the bit error rate of received samples. For a detailed discussion of such communication algorithms please refer to [300]. Finally, the decoded bits are de-interleaved and the Reed–Solomon decoder is used to reduce burst errors.

Within the RCA several critical paths exist that are subject to evaluation and are essential for a successful operation of the complete system. The critical path CP_{PHY} defines a task chain which starts at the RF frontend and lasts till the received data bits are extracted from the Reed–Solomon decoder. To prevent buffer overflows, the throughput has to match the data rate. Depending on the application, e.g., voice or data communication, the latency of the complete processing needs to be constrained. As the determined latency is in the range of a few microseconds it is far below critical values and, therefore, determined but not considered critical

The other depicted critical paths define feedback loops within the algorithm. The critical path CP_{AGC} is characterized by tight timing constraints because it is located at the RF frontend. The second depicted feedback loop determines the phase estimation loop CP_{SYNC}, which constantly needs to update the phase correction task.

Within the following case study, the tasks are shortened as Phase Correction (`PC`), Soft Demapper (`SD`), Soft Mapper (`SM`), Phase Estimation (`PE`), Soft Viterbi (`SV`), Deinterleaver (`DI`), and Reed–Solomon (`RS`).

Combination of Algorithms

The underlying hardware platform needs to be shared to execute both physical layer algorithms jointly. As both applications target different data rates, a combined execution requires more in depth analysis to find a suitable schedule. While the MIL standard processes 4,800 samples per second, the RCA runs at a higher data rate of 5,000,000 samples per second. Accordingly, the MIL standard processes one iteration while the RCA iterates 1041.7 times, which equals approximately eight frames. Therefore, the schedule combines the RCA on a frame basis and the MIL on a sample basis.

8.2.2 Overview of Processing Element

Short time-to-market and the particular pressure on low development and manufacturing costs are two of the dominating factors in today's hardware platform designs. Driven by these demands, the principle of platform- and CbD has been invented to simplify platform development. Following this paradigm a set of carefully selected IP components has been chosen for this case study. These components cover the most prominent classes of components found in today's heterogeneous MP-SoCs, among them RISC and DSP based processor cores, hardware accelerators and various communication architectures. The following listing highlights these components.

- *Generic Processor Element.* This processing element supports simplified evaluation of one or the other execution behavior by mapping arbitrary execution characteristic estimates to the element.
- *Tensilica Diamond cores.* The Diamond cores are a set of configurable processor cores that allow customization to arbitrary applications [301]. As determining an optimal processor core is not within the scope of this case study, a standard mid-range configuration has been selected throughout the complete case study. The configuration extends the base processor with floating point and DSP units. The maximum clock frequency of the inspected configuration is considered to be 500 MHz.
- *Texas Instrument C55x DSP.* The power efficient fixed point C55x DSP [290] is designed for mobile embedded devices and is considered to achieve a maximum clock frequency of 300 MHz for the purpose of this case study.
- *Texas Instrument C64x DSP.* This high performance fixed point DSP of TI [291] is designed for the needs of modern multimedia and wireless communication applications. The specification gives a maximum clock frequency of up to 1.2 GHz, but for battery powered devices a reduced a frequency is highly beneficial. Hence, within the case study the selected core is limited to a maximum clock frequency of 400 MHz.
- *Radio Frequency Hardware Accelerators.* The specially tailored RF specific hardware accelerators have been manually developed and have been characterized by fixed and deterministic latency and throughput behavior.
- *Frame and Time Synchronization Hardware Accelerators.* The accelerators have been developed for the purpose of frame and time synchronization. Again these are fully characterized by a fixed and deterministic behavior that can be extracted from the specification document.
- *Texas Instrument Viterbi Coprocessor.* Due to the extensive use of Viterbi [302] and Turbo [303] decoding, specialized coprocessors are increasingly applied for

these tasks. One prominent example is TI's Viterbi Coprocessor (VCP) Version 2 [304] that supports various modes and code rates. In addition it can operate on hard and soft decisions, making it a suitable choice for future SDRs.

- *Point-to-Point Based Communication Architectures.* Well known for long time, point-to-point communication architectures were replaced by more advanced bus or crossbar architectures. However, especially when demand for high data rates exists and the applications are well known, these architectures are utilized.
- *Bus-based Communication Architectures.* Standard bus-based communication architectures, e.g., AMBA AHB [34] or IBM CoreConnect [35] bus, can be found in nearly all MPSoC platforms. However, modern system architectures include more advanced multilayer buses, crossbar architectures, and/or even Networks-on-Chips. Since the investigated applications of the case study do not need to incorporate such complex architectures, only standard bus-based communication architectures are included within the case study.
- *Memory Architectures.* In general, memory architectures summarize different implementation options, like caches, tightly coupled memories, and shared external memories. Hence, memories are not restricted in type and size.

Please note that the case study focuses on the DSP part. Therefore, the investigated applications start after the analog–digital converter and do not include any analog components.

8.2.3 Exploration

After the introduction of the case study's key objective and starting point, the proposed exploration workflow is exemplified next. The case study follows the design process depicted in Fig. 5.1. It starts from the analytical implementation model and aims at utilizing the proposed iterative workflow for the identification and implementation of a suitable hardware design and the application-to-architecture mapping. Inherently, the temporal and spatial task mapping and the identification of task-level parallelism shall be done during the design process.

Entering this design process from the perspective of SDRs would lead to a complete software implementation of the applications, either on a DSP or general purpose processor. Since the data rates within the RCA close to the radio frontend are very high, an initial calculation is performed to evaluate the design options for this part of the application.

- Symbols having an I- and Q-value with 16 bit each (= 32 bit) arrive at a data rate of $R = 40$ Msamples/s at the RF frontend. Accordingly, the sample time is $\Delta t_{\text{sample}} = 25$ ns/sample.
- Unfortunately, this time is mostly too short for a general software implementation (5 cycles at 200 MHz clock frequency). Hence, these tasks are directly mapped onto hardwired accelerators. They have a fixed latency and throughput, thus the critical path CP_{AGC} has been verified once and does not need further analysis in the following.

In the subsequent case study these hardware accelerators are further referred to as accelerators for RF as well as frame and time synchronization. These accelerators only affect the latency calculation of the receiving critical path CP_{PHY} within the RCA by a fixed offset. In addition, the design process is entered with the following assumption of initial knowledge and software implementations available.

- A floating point reference implementation exists for the low data rate MIL-STD-188-110B communication standard. As its complexity is manageable, and the porting effort to a fixed point implementation is considered to be high [44], this reference implementation is retained.
- No reference implementation of the RCA communication standard preexists and implementation knowledge differs significantly for the application tasks.
  - The hardwired IP components for frame and time synchronization are available and an exact characterization can be given by a fixed and deterministic behavior.
  - Operation-based estimates exist for phase correction and estimation, soft de-mapper and mapper, and de-interleaver. These estimates derive from the necessary algorithmic operations.
  - The fundamental algorithms of soft Viterbi and Reed–Solomon decoder are known, but only from the algorithmic point of view.

These prerequisites define a scenario as illustrated in Fig. 8.7 with diverse implementation knowledge at the start of the design cycle. The preexisting implementation knowledge is further enhanced and refined within the case study to reveal suitable implementation candidates. Inherently, the envisioned design process and methodology are demonstrated.

The *valid range*, where the investigated system meets the latency and throughput requirements, is marked with a *gray background* for enhanced visualization throughout the different steps of the case study.

Step 1 (Analytical Model): Initial Setup

Entering the design process, the initial MPSoC platform is based on an educated guess and prior experiences with other wireless communication devices like TI's OMAP platform. Accordingly, the platform is assembled with two processing elements, one Tensilica Diamond processor core and another generic processing

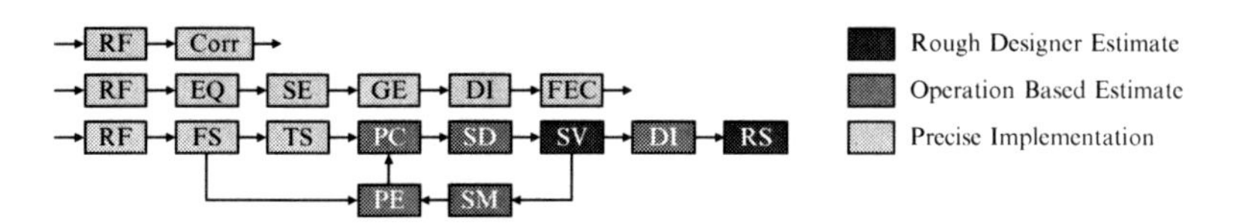

Fig. 8.7 Scenario at the initial design entry

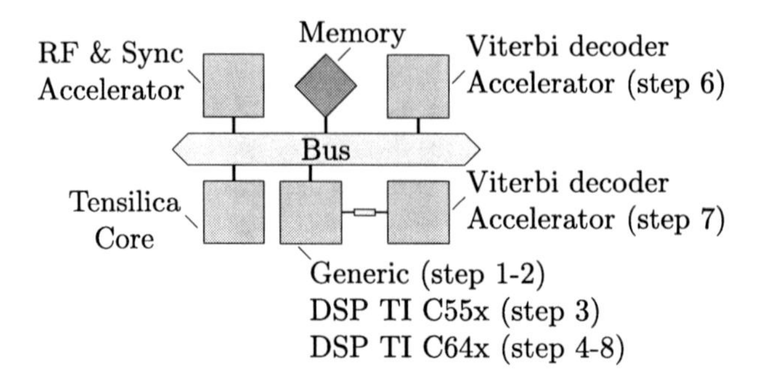

Fig. 8.8 The explored design options of the hardware architecture during the case study

element (Fig. 8.8). As a floating point software implementation of the MIL standard is available, it should be reused due to the high porting effort from floating to fixed point. As the Tensilica core is equipped with a floating point support, the MIL standard is efficiently mapped to it. The representative communication algorithm is mapped to the generic processing element to distribute the workload. Both cores are assumed to execute at a clock frequency of 300 MHz and scheduling is fixed for the specific mode and processor core to

Tensilica (MIL correlation):	`Corr`
Tensilica (MIL standard):	`EQ`, `SE`, `GE`, `MDI`, `FEC`
Generic Processor Element:	`PC`, `SD`, `SV`, `SM`, `PE`, `DI`, `RS`

Based on educated guesses, the critical paths as computed by the analytical method (Chap. 6) are depicted in Fig. 8.9. With its low complexity demands the MIL standard obviously achieves the necessary requirements successfully. In contrast, the RCA misses the constraints by far, as the computed sample processing time and feedback are significantly higher than the 25.6 µs threshold. Because of the single-core schedules, the latency shows the same characteristic as the sample processing time, hence, it is not depicted here.

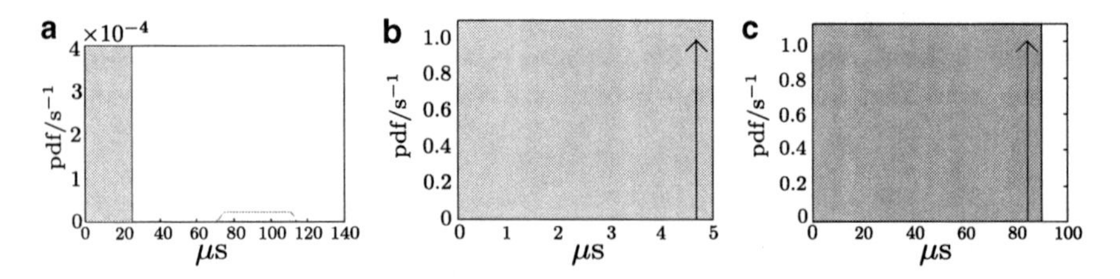

Fig. 8.9 Results for the initial setup [valid range – *gray*] (step 1). (**a**) RCA sample processing time. (**b**) MIL correlation mode sample processing time. (**c**) MIL normal mode sample processing time

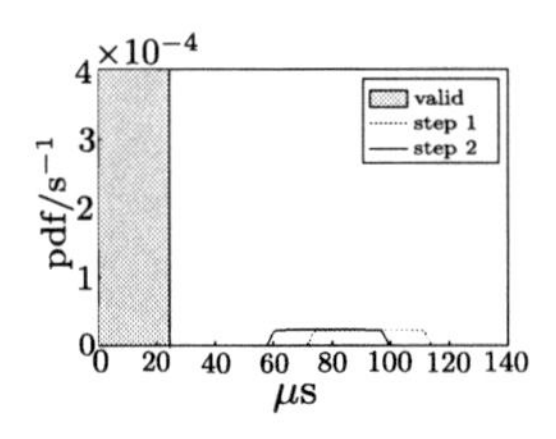

Fig. 8.10 Lowered RCA sample processing time by re-scheduling [valid range – *gray*] (step 2)

Step 2 (Analytical Model): Single Processor Core Schedule Effects

The static scheduling applied to the generic processor element needs to iterate completely to emit data from the Reed–Solomon decoder at the output. To reduce the latency and sample processing time, a slightly modified schedule enhances the processing as the Reed–Solomon Decoder finishes earlier. Nevertheless, the results (Fig. 8.10) illustrate that there is still a large performance gap for a successful implementation.

Generic Processor Element: `PC, SD, SV, DI, RS, SM, PE`

Step 3 (Analytical Model): Replacement of Generic PE by TI C55x DSP

Based on the first two analytical exploration steps, the RCA has been identified to be unsuitable for the targeted system implementation. Additionally, the major uncertainties are collocated within the soft Viterbi and Reed–Solomon decoder of the addressed implementation. Apparently, modifications of the implementation should provide more precise estimates for these two relevant tasks. These two tasks are well-known in the domain of DSP and can be efficiently implemented on available DSP architectures. Since energy efficiency is a dominant issue in designing wireless communication devices, first the impact of the TI C55x DSP is evaluated. The implementation of both tasks is precisely defined by a vendor-provided software IP library [305] (Fig. 8.11). Annotating their precise characteristics and updating the annotation characteristics of the other tasks (Fig. 8.11) computes the analysis results illustrated in Fig. 8.12.

These results demonstrate that the previous coarse-grained estimates have been too optimistic, as visible in the increasing average RCA sample processing time. Unfortunately, this is common or at least often occurs in early design stages. This issue is one of the main reasons for the usage of the analytical implementation model, because evaluation of a single design point requires only minutes.

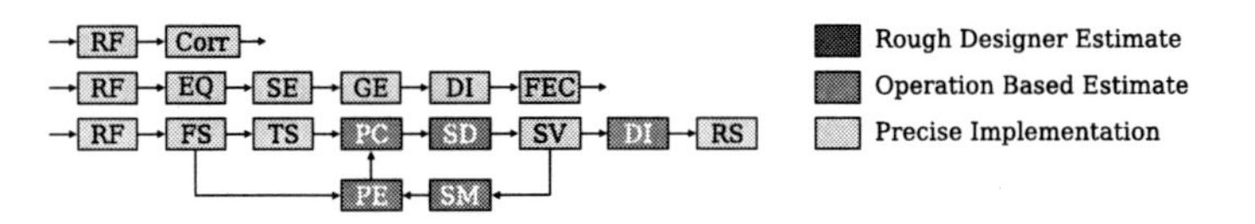

Fig. 8.11 Implementation knowledge after refinement to TI C55x DSP (step 3)

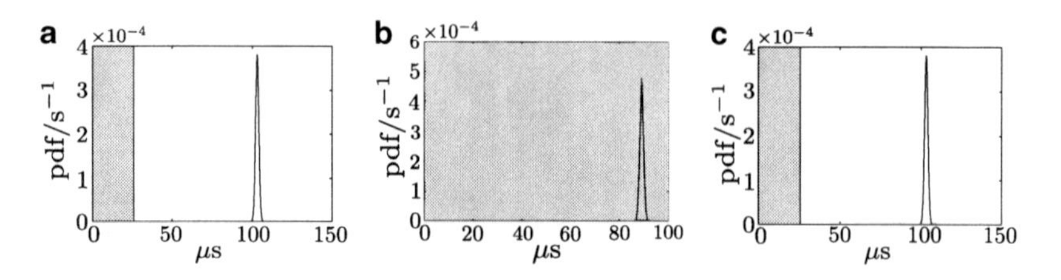

Fig. 8.12 Results for the architecture refinement: C55x DSP [valid range – *gray*] (step 3). (**a**) RCA sample processing time. (**b**) RCA latency (not critical). (**c**) RCA feedback delay

Thanks to the gained implementation knowledge, more precise input characterization can be given. Naturally, this significantly reduces uncertainties in the results, but the application requirements are still not met. In a deeper investigation, the Reed–Solomon decoder has been identified to be the bottleneck as execution takes far too long. Hence, the next design step optimizes this task to speed-up the overall execution.

Step 4 (Analytical Model): Replacement of TI C55x with C64x DSP

With the given design restrictions and a rather pragmatic design approach, the TI C55x DSP is replaced by the much more advanced and powerful TI C64x DSP. Especially, the Galois field instructions [306] of the TI C64x DSP accelerates the execution of the Reed–Solomon decoder [305] by approximately one order of magnitude. In addition, the typical clock frequency is higher than the 300 MHz and is assumed to be 400 MHz for the given scenario.

Illustrated in Fig. 8.13, the analysis results demonstrate that the RCA constraints are still not met, although a huge performance gain is achieved. However, the performance is close to the constraint threshold and the requirements might be kept with only minor system modifications. Since the processor cores execute at 300 MHz (Tensilica core) and 400 MHz (TI C64x DSP), a simple solution would increase the frequency to accelerate the overall execution. This increase would definitely

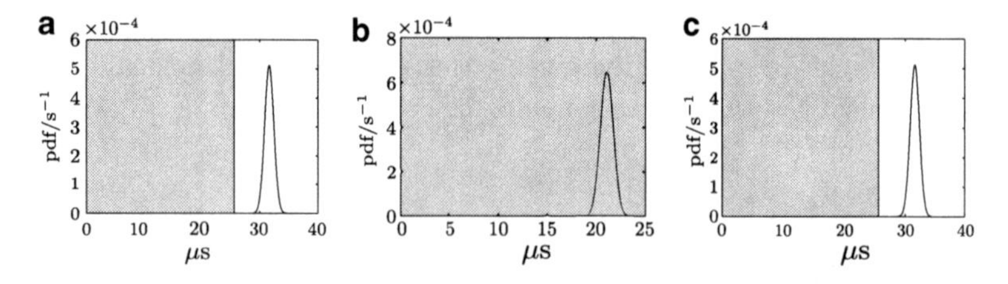

Fig. 8.13 Results for the architecture refinement: TI C55x DSP to C64x DSP [valid range – *gray*] (step 4). (**a**) RCA sample processing time. (**b**) RCA latency (not critical). (**c**) RCA feedback delay

result in an increase in the power and, hence, energy consumption. Apparently, more innovative solutions should be added to minimize the necessary clock frequency increment. For example, one part is the rescheduling of the task execution discussed next.

Step 5 (Analytical Model): Task Rescheduling

To accelerate the execution of both communication standards a joint schedule is envisioned, which interleaves the MIL and RCA execution. Since the FEC decoder of the MIL standard can be efficiently implemented on the DSP, a modified mapping sorts the FEC decoder to the DSP while the annotated performance is profiled by instruction set simulation. Conversely, the Tensilica processor core can relieve the DSP from performing the PC, SD, SM, and PE tasks. These tasks are not too computationally intensive and the assumption is made that they execute similarly on the Tensilica processor core.

The schedules are separated into two parts as eight iterations of the RCA are executed within the time of one MIL standard iteration. Hence, the schedules are as follows.

Tensilica(0):	`EQ, SE, GE, MDI`
C64x(0):	`PC0, SD0, SV0, SM0, PE0, DI0, RS0, FEC0`
Tensilica(1-7):	`PCi, SDi, SMi, PEi`
C64x(1-7):	`SVi, DIi, RSi, FECi`

Furthermore, the clock frequency of the Tensilica processor is moderately increased to 500 MHz to speed up the execution.

The computed results are depicted in Fig. 8.14. Due to the interleaved scheduling, the execution characteristic differs over the iterations of the high data rate RCA. Since eight RCA iterations compute within the iteration time of the MIL standard, the second and the last RCA iterations differ from the other iterations. Under the assumption of sufficient buffer capacities available, the aggregated maximum sample processing time of eight iterations, given by the real-time constraints, equals $8 \times 25.6\mu s = 204.8\mu s$.

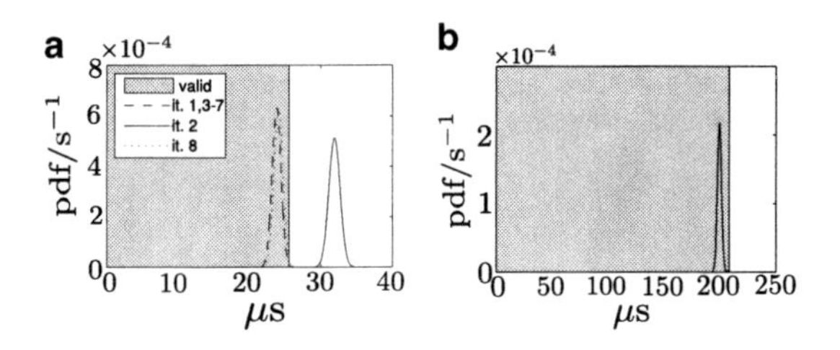

Fig. 8.14 RCA results for interleaved schedule and increased clock frequency [valid range – *gray*] (step 5). (**a**) Sample processing time. (**b**) Agg. sample processing time

While the constraint is not met when investigating every single iteration, the aggregated sample processing time looks promising to keep the aggregated constraint. Accordingly, a first implementation candidate has been identified in the analytical implementation model within merely five iteration steps, with each step requiring only minutes for computation of the analysis results.

Step 6 (Abstract Simulation Model): Validation of the Implementation Candidate

One central element of the proposed workbench and methodology is the smooth transition from one abstraction level, either analytical or simulation-based, to the other. To demonstrate this and to prove the capabilities of the approach, the found implementation candidate is evaluated within the abstract simulation implementation model. The advantage of the simulation environment is that dynamic effects like bus contentions are explicitly modeled and are incorporated within the obtained simulation results.

The abstract simulation model is composed by utilizing the VPU and Time Retrieval Engine (Chap. 7), and simulation results are determined as illustrated in Fig. 8.15. These results show only minor deviation from the previous analysis results. Based on this investigation, an increasing confidence for the feasibility of the implementation candidate is obtained, as only limited bus contentions appear and other dynamic effects do not significantly impact the targeted implementation.

Since there is no information that requires back-annotation, other implementation candidates can be evaluated before moving toward the final implementation or, instruction set simulation can be targeted directly. For the exemplary case study, the first option to identify more feasible implementations is chosen.

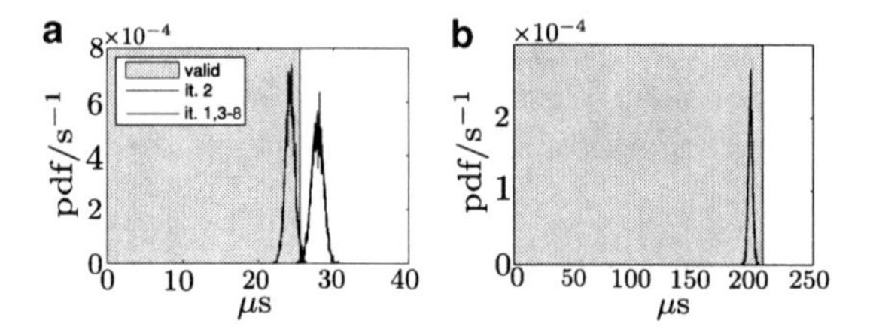

Fig. 8.15 Simulation results for the aggregated RCA sample processing time [valid range – *gray*] (step 6). (**a**) Sample processing time. (**b**) Agg. sample processing time

Step 7 (Analytical Model): Adding a Viterbi CoProcessor

As commonly applied in today's wireless communication devices, a Viterbi CoProcessor (VCP) is included to speed-up the execution of this specific task. For the

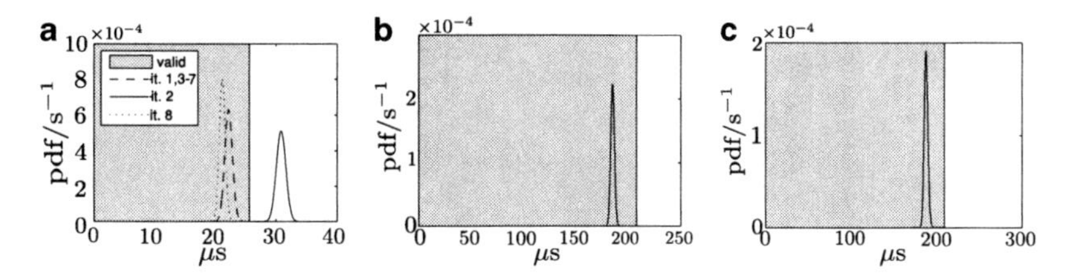

Fig. 8.16 Results for system including VCP connected to a bus architecture [valid range – *gray*] (step 7). (**a**) RCA sample processing time per sample. (**b**) RCA aggregated sample processing time. (**c**) MIL sample processing time

inspected scenario, the VCP runs at the third clock frequency of the DSP. In the following, two design options are investigated. While step 7 inspects the VCP connected to the bus-based communication architecture (Fig. 8.8), step 8 investigates the VCP tightly coupled to the DSP by a FIFO communication channel.

Connecting the VCP to the bus communication architecture, additional delay is incurred because of the communication overhead. However, the analysis results (Fig. 8.16) show a speed-up by parallelism that compensates the communication overhead. Hence an overall reduced sample processing time is measured and the configuration found is a feasible implementation candidate.

Step 8 (Analytical Model): Tightly Coupled Viterbi CoProcessor

For lowering the encountered communication overhead, the VCP is directly coupled with the DSP. Unfortunately, this impacts the scheduling so that the proceeding and following tasks of the Viterbi task need to be executed on the DSP. Otherwise the DSP has to forward the data or the spatial task mapping is invalid. For the given example this requires re-scheduling, breaking the optimized scheduling for the given application.

Tensilica(0):	`EQ`, `SE`, `GE`, `MDI`
C64x(0):	`PC0`, `SD0`, `SV0`, `SM0`, `PE0`, `DI0`, `RS0`, `FEC0`
Tensilica(1-7):	`PCi`, `PEi`
C64x(1-7):	`FECi`, `SDi`, `SMi`, `DIi`, `RSi`,
VCP(0-7):	`SVi`

The obtained results are illustrated in Fig. 8.17 depicting another possible implementation candidate. It has to be noted that this system shows worse results than the previous one, but the Tensilica processor core is less utilized. This might result in a trade-off decision in case another application needs to be supported later on. Therefore, it is hard to judge whether one or the other implementation is better.

To demonstrate the interwoven complexity, the same schedule has been additionally implemented for the bus-based communication VCP of step 7 (Fig. 8.18).

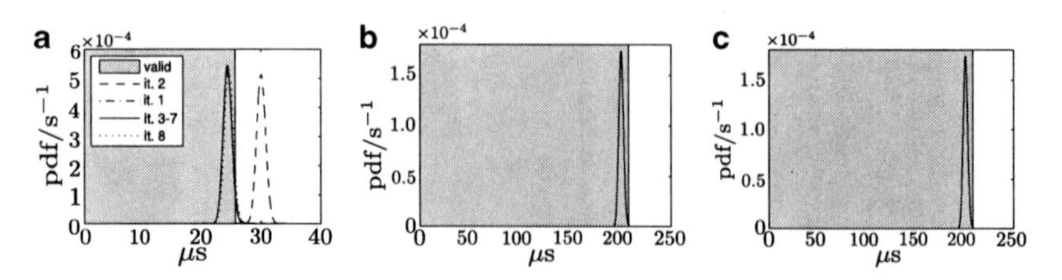

Fig. 8.17 Results for the tightly coupled VCP [valid range – *gray*] (step 8.1). (**a**) RCA sample processing time per sample. (**b**) RCA aggregated sample processing time. (**c**) MIL sample processing time

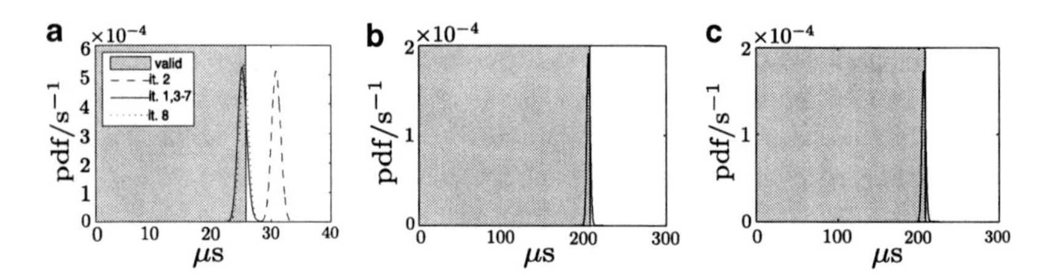

Fig. 8.18 Results for the bus connected VCP with unoptimized scheduling [valid range – *gray*] (step 8.2). (**a**) RCA sample processing time per sample. (**b**) RCA aggregated sample processing time. (**c**) MIL sample processing time

The obtained analysis results illustrate the trade-off decision between the optimized schedule and selection of the communication architecture. As visible, the impact of the schedule is larger than moving parts of the data communication to a dedicated point-to-point communication architecture for the inspected scenario.[2] Please note that there might be other schedules which would better suit the tightly coupled VCP and finally exploit the optimized communication architecture efficiently in a system-wide context.

The earlier encountered issue nicely demonstrates the need for a system-wide investigation as the individual design decisions, schedule, and point-to-point communication architecture, conflict with each other and an optimal trade-off decision must be taken.

Further Exploration Steps and Process

Three possible implementation candidates have been identified. In addition, the investigation of the analytical model and the abstract simulation model has briefly

[2] Please note that this result is only valid for the given scenario. In general, the point-to-point communication architecture should be able to achieve superior performance.

outlined the capabilities of the proposed methodology by simply going up and down the abstraction levels. The next step is to replace the VPU model for at least one of the evaluated candidates by instruction set simulators.

It might be noted that during this refinement results obtained earlier might become invalid because of the encountered dynamic effects. These need and can be integrated by back annotation, which allows comparing implementation options at higher abstraction layers, thus speeding up the exploration and evaluation process.

8.3 Summary of the Case Study

The case study and results have demonstrated the flexibility and usefulness of the proposed design framework and methodology. First, the accuracy of the annotation techniques has been inspected. In comparison to cycle accurate instruction set simulation, the measured annotation errors show only minor deviation, which is adequate for early design space exploration. In addition, investigations of different hardware and software options have emphasized the requirements for multiple annotation techniques based on various fundamental concepts and methodologies. Derived from this and prior experiences, a recommendation table (Table 8.2) has been extracted that helps system architects to select the best possible annotation with respect to the targeted scenario.

The second part of the case study has highlighted the envisioned design process based on a practical SDR design. This case study has studied the interwoven design decisions on the system level. The analytical implementation model served as a key methodology for identification of feasible design candidates. Additionally, the abstract simulation model based on the VPU has identified and validated dynamic effects within anticipated implementation candidates. The case study finished with three identified implementation candidates for future implementation and refinement to instruction set simulation. Typical trade-off decisions – such as task re-scheduling and selection of processing elements as well as communication architectures – have been encountered and evaluated.

Chapter 9
Summary and Outlook

The demand of end consumers on product and technology advances puts high pressure on the design of next generation mobile wireless communication devices. Especially, the crisis of complexity [5] and the pessimistic voices predicting the end of Moore's Law and Dennard's Scaling rule prohibit simply relying on technology shrinks. System architects are more than ever being expected to provide innovative designs to achieve the given requirements in order to cope with the expectations of end users.

Multi- and upcoming many-core architectures are currently considered to be the optimal choice to increase the performance of such platforms. However, these impose other design challenges, mainly due to the parallel processing. From a theoretical point of view, doubling the number of processor cores should in general double the performance. Unfortunately, in a practical scenario this typically does not apply by far. Issues of nonoptimal application partitioning, as well as communication and synchronization are responsible for this expected and actual performance mismatch.

Supplementary to this issue, battery capacities have not been significantly increased in the past, in contrast to the complexity of applications and hardware platforms. This has led to an increasing performance-energy gap which demands hardware platforms to boost performance and energy efficiency. As a consequence system architects are applying increasingly heterogeneous Multi-Processor System-on-Chip solutions to meet the stringent performance and energy demands.

Besides these pure computational centric considerations, communication architectures and memories play an extremely important role in today's designs. For example, the memory subsystem commonly occupies a large portion of the chip area and communication architectures are responsible for a significant amount of the energy consumption. Therefore, the design process has to consider the complete hardware architecture and not just single components.

In addition, the mapping of applications to the various processing elements of a heterogeneous MPSoC platform defines a fundamental design decision for a successful system implementation. Especially in the domain of wireless communication, the tight real-time performance constraints makes the mapping step highly challenging, but has significant impact on the system's performance. This mapping comprises both temporal and spatial task mapping. Here, temporal mapping defines

T. Kempf et al., *Multiprocessor Systems on Chip: Design Space Exploration*,
DOI 10.1007/978-1-4419-8153-0_9, © Springer Science+Business Media, LLC 2011

the execution order of tasks onto a single processing element, while the spatial mapping denotes the task execution on a particular processing element. Inherently, the mapping restricts, or at least influences the software development process, as the software development technique varies depending on the underlying architecture.

Altogether system architects are facing many design decisions and options. These need to be selected optimally, since the rapidly moving markets in the domain of wireless communication do not forgive wrong design decisions. Today these decisions are more likely taken on an ad-hoc basis and based on previous designs rather than being based on a sophisticated analysis. The increasing complexity of platforms and the resulting high risks of false or suboptimal decisions, demand new and enhanced design methods to guide designers right from the start of the design cycle to the final implementation.

Such design methodology has been demonstrated within this book. In contrast to the commonly used approaches based on outdated Excel sheets and/or Gantt charts, the proposed methodology defines a structured design process starting at a high abstraction level leading toward the final implementation. Besides the concept itself, the design methodology has been demonstrated by a tool environment covering the envisioned design process. Already today parts of the environment have been included into commercialized tools and have successfully found adoption in industrial projects [171] (Appendix C).

The key contribution of this book is the early design space exploration framework as a whole. The natural entry point for the design process is a coarse-grained description and characterization of both application and architecture. The key idea of the proposed design process is that exploration and evaluation of a target system can be performed on a set of different abstraction levels. Starting at the highest abstraction level, evaluation of the complete system takes only seconds. In addition, the framework inherently supports a smooth transition from high to low level of abstraction. Once a suitable implementation candidate has been identified, this candidate can be quickly investigated at the lower abstraction level. Possible identified issues and mismatches can be easily backpropagated to the higher levels and exploration of various design decisions can be performed efficiently at the higher abstraction layer.

In addition to the introduced design space exploration framework, other main contributions are the abstraction layers above instruction set simulation that can be separated into the following.

- At the highest abstraction level the *analytical implementation model* computes the system performance based on the individual execution characteristic of the application tasks that are mapped onto a given implementation candidate. Prerequisites for this calculation are the descriptions of the individual task execution characteristics by means of random variables that define the execution behavior. The absence of any implementation centric model, e.g., software implementation or detailed hardware model, allows applying this exploration technique prior to the implementation process within the conceptional phase.
- Still on the conceptional level, the *abstract simulation implementation model* enables investigation of more detailed hardware and software effects. In addition,

the mixing of detailed implementation models with abstracted representations is possible and a clear link to the final implementation exists. Main enabler is the *VPU* with various modeling and annotation techniques to capture the multiple hardware and software implementations.

Based on the smooth transition between the abstraction levels and the simple definition of different scenarios, a large number of configurations can be explored especially when utilizing the analytical model. For acceptance and practical use within designs, the framework has been integrated directly from the start into commercially available tools following a library concept. For the statistical analysis the favorite tool is MATLAB, while for system simulation SystemC as language and CoWare Platform Architect as tool have been selected. In spite of the current implementation, the framework is not limited to these tools. The clear benefit of this integration is the capabilities to analyze the evaluated system in visualized formats provided by these environments. The graphical capabilities of MATLAB allows efficient evaluation of the analytical model, while the simulation-based environment inherently supports arbitrary tracing facilities like message sequence charts or specific latency and throughput measurements based on CoWare's Analysis technology.

The proposed framework has been validated by a case study consisting of two major parts. The first part of the case study discusses the accuracy of the proposed annotation techniques. The annotations of different implementation options – like general purpose processors with software developed in generic high-level languages and DSPs with low-level software implementations – have quantified the incurred modeling errors. Comparisons to cycle accurate simulation models have illustrated the precise modeling capabilities for the purpose of early design space exploration.

The second part of the discussed case study proves the usability and exploration capabilities of the introduced framework. The inspected scenario defines a general design problem from the domain of wireless communication and SDRs. The design focuses on the development of a receiver platform for two given standards, a high and low data rate one. Both of these standards need to execute in parallel on a shared hardware platform. Additionally, the limiting constraints of development effort and costs restrict the design to mainly standard IP components together with application specific IP components. This design exemplifies the use of the proposed framework and highlights its capabilities.

Entering the design process with an initial guess, hardware and software development of this complex scenario starts at the analytical implementation model. The fast exploration capabilities at this level allow excluding nonsuitable design options and efficiently identifying promising ones. In an iterative design loop, the candidates are further evaluated and different implementations are explored till a final implementation candidate is identified.

Outlook

The increasing complexity issues of upcoming multi- and many-core architectures are manifold and definitely not all to be solved by the design space exploration framework. The proposed methodology can only serve as an entry point which allows efficient exploration of various design decisions. The following highlights ideas and other prospects for future research.

- *Further case studies and more complex scenarios* including latest processor cores such as the ARM Cortex-A8 as well as commercially available standards would guarantee reliability and confidence in the proposed methodology. Possible applications can not only be based on existing standards such as WLAN or WIMAX but also on still developing ones like LTE.
- *Separation of the annotation characteristics* into a software and hardware characterization would lead to increasing efficiency for the modeling and evaluation of different use cases. This large scale research project should aim at the identification of common hardware features and a condensed representation of these for a particular processor core. Finally, this might end in a library in which each particular processor core is stored in terms of a suitable characterization. At runtime, both application and architecture are considered and at run-time, computed characteristics reflect the final implementation.
- A suitable extension of the VPU should target *upcoming Multi-Processor (MP) cores* like the ARM Cortex-A9. As the current realization is limited to single core modeling, such multicore scenarios must be modeled with each included core as a VPU. This makes the modeling of task migration challenging. For this purpose the VPU simulation model needs to incorporate capabilities to mimic the behavior of the multiple cores within.
- Another major challenge for future multi- and, especially, many-core platforms will be the *resource management*. This will become, at least from our perspective, one of the major issues in future designs. Due to the significant performance impact, this issue needs an efficient methodology to determine optimal management strategies at the very beginning of the design cycle. In close cooperation with the VPU as a vehicle for early design space exploration, a unique and fundamental combined exploration technique should be envisioned.

Appendix A
Advanced Features of the Analysis Framework

A.1 Analysis Graph Simplification

The complexity of the intended applications and application task graphs can be assumed to be within the range of hundreds to thousands of tasks. With respect to this, graph size simplifications and complexity reduction is mandatory for the analysis algorithm. Addressing this issue, the analysis algorithm implementation incorporates two key techniques. In addition, a looped application should be considered so that further simplification is achieved by reapplying the other technique.

A.1.1 Task Merging

Adjacent graph vertices (tasks) are merged if and only if the following conditions hold.

- The first vertex in a sequence is a join or has a predecessor being a start, end, or way-point of a critical path, but is no split vertex.
- The last vertex in a sequence is a split or has a the successor being a start, end, or way-point of a critical path, but is no join vertex.
- Other vertices in the sequence must not be a join, split, or way-point of a critical path.

For the example depicted in Fig. A.1a the number of nodes can be reduced by vertex merge operation from $|V| = 71, |E| = 93$ to $|V| = 41, |E| = 63$.

A.1.2 Shortcut Elimination

Redundant edges are typically avoided in the original task graph. However, after simplifications, these edges occur as illustrated in Fig. A.1b for example. A redundant edge is given by an edge (v_t, v_h) so that there is another path from v_t to v_h within

T. Kempf et al., *Multiprocessor Systems on Chip: Design Space Exploration*,
DOI 10.1007/978-1-4419-8153-0_10,

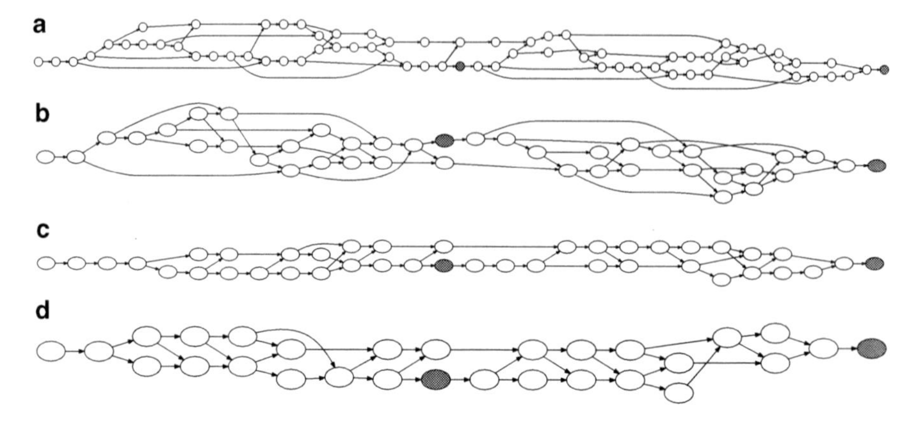

Fig. A.1 Exemplified analysis graph simplifications. (**a**) Original Analysis Graph as constructed from DFG and CFG. (**b**) Merging of nodes reduces the number of vertices. (**c**) Removal of redundant edges. (**d**) Further merging of nodes reduces the number of vertices

the Analysis Graph (AG). As the longer path limits the overall execution, the edge is inherently captured within the other path and hence is redundant as an alternative path exists. Figure A.1c illustrates the impact on the AG by the removal of these shortcuts. For the given example, the number of edges is reduced from $|E| = 63$ to $|E| = 53$ (Fig. A.1b).

A.1.3 Iterative Application

An alternating application of both simplification techniques leads to optimal results in terms of complexity reduction. After shortcut elimination, the task merging can further reduce the AG complexity from $|V| = 41, |E| = 53$ to $|V| = 29, |E| = 41$ for the given example highlighted in Fig. A.1d.

A.2 Scheduling Scenarios

Selecting one or the other scheduling can have tremendous impact on the overall system performance. Accordingly this topic has attracted many researchers. The framework operates on the foundation of a workbench, leaving the identification of the optimal schedule to the designer or other tools. However, the analysis allows for evaluation of the performance impact of selected schedules and provides an efficient technique for the specification of such.

The following discussion demonstrates the possible performance impact and the provided evaluation and specification technique of the framework based on an exemplary scenario.

With the system scenario specified in Fig. 6.8, the temporal and spatial task mapping is defined as the two schedules of the processing elements pe_a and pe_b.

$$SC(pe_a) = (T_1, T_2, T_5, T_7)$$
$$SC(pe_b) = (T_3, T_4, T_6, T_8)$$

Starting with the initial temporal and spatial task mapping, the question arises how to replace the task execution T_5 with the required two instances $T_{5,0}$ and $T_{5,1}$. Obviously, different solutions exist which will impact the overall performance.

Following a pragmatic solution, the required two task instances replace the task T_5 within the schedule of pe_a, so that the resulting schedule is given by the schedule depicted in Fig. A.2 which equals

$$SC(pe_a) = (T_1, T_2, T_{5,0}, T_{5,1}, T_7).$$

Based on the given task's timing assumptions, significant stall periods are encountered leading to processing elements with large intervals of idle time and a long overall execution time. Simply rearranging the schedule and moving the execution of task T_2 within the two iterations of the task $T_{5,0}$ and $T_{5,1}$ achieves superior performance and shortens the overall execution time (Fig. A.2).

$$SC(pe_a) = (T_1, T_{5,0}, T_2, T_{5,1}, T_7)$$
$$SC(pe_b) = (T_3, T_{4,0}, T_{4,1}, T_{6,0}, T_{6,1}, T_8).$$

Furthermore, different iterations of the complete schedule can interleave so that tasks of different iterations execute simultaneously. For the given example this is

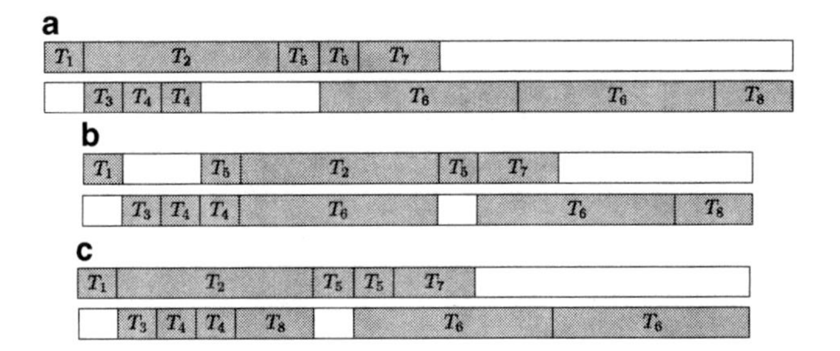

Fig. A.2 Exemplary schedules. The upper chart pictures $SC(PE_A)$ and the lower one $SC(PE_B)$. (**a**) Initial schedule based on the topological task sequence of the initial task graph. (**b**) Schedule modification based on task T_5 instances. (**c**) Task scheduling with interleaved iterations

depicted in Fig. A.2. Here task T_8 operates on the date of the previous iterations of task $T_{6,0}$, $T_{6,1}$, and T_8. Interleaving the schedule introduces a typical trade-off decision between latency and throughput, formulated as follows.

- An increased throughput (equals a decreased scheduling time Δt) is achieved due to a shorter length of the schedule per iteration. For the given example schedule length decreases from Δt to $\Delta t'$.

$$\begin{aligned}\Delta t &= \Delta t(T_2) + \Delta t(T_5) + \Delta t(T_6) + \Delta t(T_6) + \Delta t(T_8)\\ \Delta t' &= \Delta t(T_3) + \Delta t(T_4) + \Delta t(T_4) + \Delta t(T_8) + \Delta t(T_5) + \Delta t(T_6) + \Delta t(T_6).\end{aligned}$$

- In contrast to the throughput, there is an increase in the latency, defined as the delay between a sample that enters the algorithm and the information that leaves the algorithm. For the given example the latency rises from L to L'.

$$\begin{aligned}L &= \Delta t + \Delta t(T_1)\\ L' &= \Delta t' + \Delta t(T_1) + \Delta t(T_3) + \Delta t(T_4) + \Delta t(T_4) + \Delta t(T_8).\end{aligned}$$

A.2.1 Scheduling Definition Within the Analysis Framework

In order to support the investigation of different temporal and spatial task mapping decisions, the framework introduces a condensed language to specify the envisioned scheduling with mixed task types. The grammar of this language is defined as follows.

$$\begin{aligned}\text{schedule} &\rightarrow \text{task} \mid \text{task}\ \mathbf{,}\ \text{schedule}\\ \text{task} &\rightarrow \text{taskname itmod blkmod}\\ \text{taskname} &\rightarrow \mathbf{identifier} \mid \underline{\text{identifier}}\ \_\ \underline{\text{instance}}\\ \text{blkmod} &\rightarrow \mathbf{!} \mid \varepsilon\\ \text{itmod} &\rightarrow \mathbf{+}\ \underline{\text{iterations}} \mid \varepsilon\end{aligned}$$

The schedules $SC(pe_a) = (T_1, T_2, T_5, T_7)$ and $SC(pe_b) = (T_3, T_4, T_6, T_8)$ define the first example (Fig. A.2). The second example requires only minor modification of the schedule $SC(pe_a) = (T_1, T_{5,0}, T_2, T_{5,1}, T_7)$ and the final step interleaving the different schedule iterations is written as $SC(pe_b) = (T_3, T_4, T_8 + 1, T_6)$.

A.3 Dependency Delays

The mathematical equation given in Sect. 6.2.2 formulates the calculation rule for a single dependency delay node. During the analysis of the complete system, several dependency delays commonly need to be calculated and an algorithm is required

Algorithm 2 Calculation of the dependency delays

Input: Analysis Graph AG
Output: Modified Analysis Graph AG

1 $V_S = \{v \in V(AG) : |\{(v, v_i) \in E(AG), v_i \in V(AG)\}| > 1\}$;
2 $V_J = \{v \in V(AG) : |\{(v_i, v) \in E(AG), v_i \in V(AG)\}| > 1\}$;
3 $\tilde{V}_J$ = filter_joins_in_critical_paths($\tilde{V}_J$);
4 $\tilde{V}_J$ = sort($\tilde{V}_J$);
5 **foreach** $v_j \in \tilde{V}_J$ **do**
6 $\quad V_{\text{predec}} = \{v \in V : \exists(v, v_j) \in E\}$;
7 $\quad V_D = \emptyset$;
8 $\quad$ **foreach** $v_p \in V_{\text{predec}}$ **do**
9 $\quad\quad$ Add vertex v_d in AG;
10 $\quad\quad E(AG) = (E(AG) \setminus (v_p, v_j)) \cup \{(v_p, v_d), (v_d, v_j)\}$;
11 $\quad\quad V_D = V_D \cup \{v_d\}$;
12 $\quad$ **end**
13 $\quad v_s$ = First common split on reverse paths;
14 $\quad$ **foreach** $v_d \in V_D$ **do**
15 $\quad\quad X_{(v_s, \ldots, v_d)}$ = calculate_performance_characteristic(v_s, v_d);
16 $\quad$ **end**
17 $\quad$ **foreach** $v_d \in V_D$ **do**
18 $\quad\quad X_m = max(\{X_{(v_s, \ldots, v_i)} : \forall v_i \in V_d \setminus \{v_d\}\})$;
19 $\quad\quad X_d = max(0, X_m - X_{P(s,i)})$;
20 $\quad$ **end**
21 **end**

to identify the necessary insertion points. Algorithm 2 creates dependency delays when necessary and calculates the corresponding execution characteristic in terms of a random variable.

A.4 Practical Calculation and Stochastic Independence

A *Monte-Carlo* methodology is introduced to evaluate the effect of neglecting the occurring stochastic dependence during calculation of critical path characteristics in the presence of data dependency delays. Additionally, the Monte-Carlo technique can be utilized when computational complexity gets too large for stochastic analysis and detailed simulation might be not available or executes significantly more slowly.

In general the Monte-Carlo technique generates N realizations of all random variables $\Delta t \sim X(task, PE)$ within a given scenario. In a second step these realizations are utilized to compute the characteristics of each critical path for the sampled values. Finally, the aggregation of all N realizations retrieves the stochastic distribution of each specified critical path. For practical considerations the number of investigated realizations N must be large.

```
1  io_routine() {
2      copy_from_user(buffer, p, count);
3      set_up_dma_controller();
4      scheduler();
5  }
6
7  isr_routine() {
8      acknowledge_interrupt();
9      unblock_user();
10     return; // from interrupt
11 }
```

Listing B.3 C-based Implementation [273]

```
1  io_routine() {
2      copy_from_user(buffer, p, count);
3      DMA_Protocol();
4      scheduler(); // OsSuspend(...)
5  }
6
7  isr_routine() {
8      acknowledge_interrupt();
9      unblock_user();  // OsResume(...)
10     return; // from interrupt
11 }
```

Listing B.4 VPU implementation

Fig. B.2 Example of DMA based I/O device driver

Appendix C
Task Modeling and Virtual Processing Unit

In 2007, the VPU and Task Modeling framework have been transferred to CoWare Inc. (recently acquired by Synopsys Inc. in 2010). After thorough evaluation of the first prototype by early access customers, the technology has been extended and integrated into Synopsys' Platform Architect tool. In the following, this Task Modeling framework and the VPU are presented briefly.

C.1 Overview

At the time of writing of this book, the commercially available framework inherits most of the concepts and technologies of the abstract simulation model described throughout Chap. 7. In the first place the technology addresses system architects. The major use-case is to evaluate whether an application can be executed on a given hardware architecture within the given performance requirements or not. These investigations need to be carried out prior to the actual development phase, because they can introduce significant design changes.

Before giving a detailed description of the commercially available framework, the requirements and goals of the approach shall be recaptured. The major focus is to identify a suitable task partitioning, mapping, scheduling, and synchronization along with an efficient task switching policy. Additionally, the dimensioning of the communication architecture and the memory subsystem are important steps within the MPSoC design addressed by the technology. For efficient design space exploration, a quick and simple mechanism for creating, simulating and analyzing different scenarios is required and the modeling accuracy must be close to the final implementation. Finally, a stepwise refinement flow from simple models to detailed ones is necessary to close the gap to existing and widely accepted evaluation techniques. These requirements have led to the current Task Modeling and VPU methodology described next.

The general design flow is illustrated in Fig. C.1. The underlying concept of tasks is essential. A task is considered as an independently running part of an application

T. Kempf et al., *Multiprocessor Systems on Chip: Design Space Exploration*,
DOI 10.1007/978-1-4419-8153-0_12,

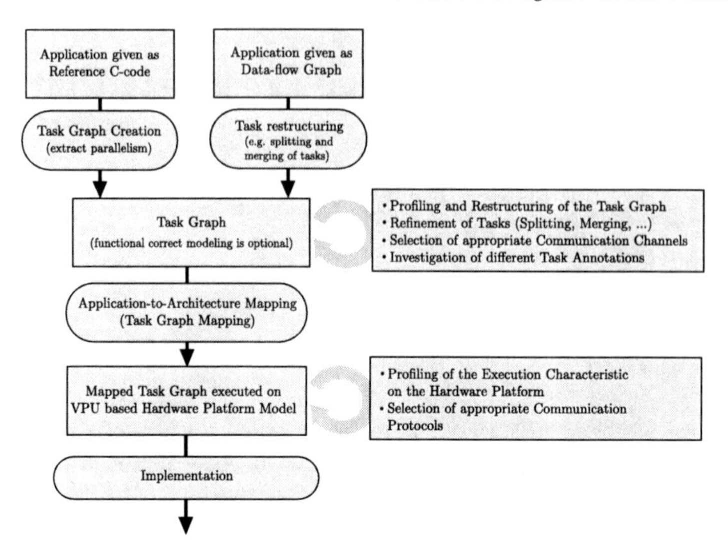

Fig. C.1 Overview of the design flow (based on [307])

that has input and output ports. Tasks can only exchange data over these inputs and outputs, which in turn are connected to communication channels. Together the tasks and channels represent the application.

The level of detail for modeling an individual task is given by the functionality and the annotation of the execution characteristic (see Fig. 7.3). Typically, at first a nonfunctional model is developed, that is based on traffic generators with statistical or coarse-grained source-level annotations of the execution characteristic (see Sect. 7.3). These annotations can be captured for example by the number of cycles required for task execution on an anticipated processing element. These numbers can either be estimates based on expert knowledge or gained by measurements. This way performance data of the complete system can be captured quickly, which provides helpful guidance during system development. Over time, the nonfunctional tasks are replaced with functionally correct tasks, enabling detailed performance analysis as required in later design stages.

In the first phase of the design flow, a task graph is constructed for each application, which can be extracted from a sequential reference code, e.g., an application given as C-program, or from a data-flow model. Nowadays, similar data-flow modeling can be found in SW development tools in the domain of DSP and microcontroller programming, TI C6Flo [308] is an example of such flow.

Having created this initial task graph, the system architect can quickly profile, restructure, and refine the tasks. This includes splitting and merging of tasks, selecting different communication channels, evaluating different processing elements by modifying the annotations of the tasks and so on.

At every point in time the task graph, whether nonfunctional or fully functional, can be mapped and profiled on a hardware platform model based on the VPU. The mapping is performed in a convenient graphical user interface by a simple drag and drop mechanism. In this phase, the load of the processing elements caused by the execution of the application is investigated. Furthermore, the effect of the communication architecture and memory subsystem can be analyzed under the given workload.

Targeting the final product, the implementation step requires the replacement of the VPU by a detailed simulation model of the processing element. This includes the cross-compilation of application tasks mapped onto the processor cores.

The following sections examine the different design phases and emphasize the inherited concepts described earlier in this book.

C.2 Task Graph Assembly and Analysis

Once all application tasks have been modeled, the complete application can be assembled and analyzed. The Task Modeling and VPU IP component are fully integrated into the Synopsys products. Therefore, developers can make use of the PCT [173] for enhanced visualization and easy creation of task graphs. Figure C.2 illustrates the task graph of a MIMO OFDM transceiver. Modeling a task graph follows the same principle as of modeling a regular SystemC hardware platform. In addition, the SystemC model creation has been enhanced by a declarative composition of task graphs. This minimizes the need for time consuming recompilation and significantly reduces the time for exploring different task graph structures. For example, the task graph can be rearranged or multiple tasks of the same type can be instantiated without recompilation.

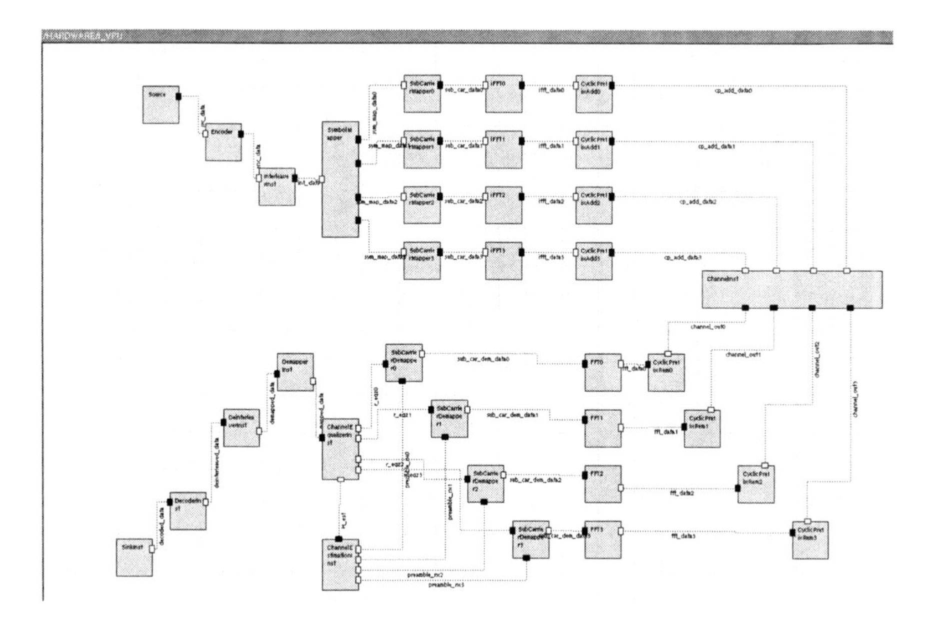

Fig. C.2 Exemplary task graph in synopsys PCT

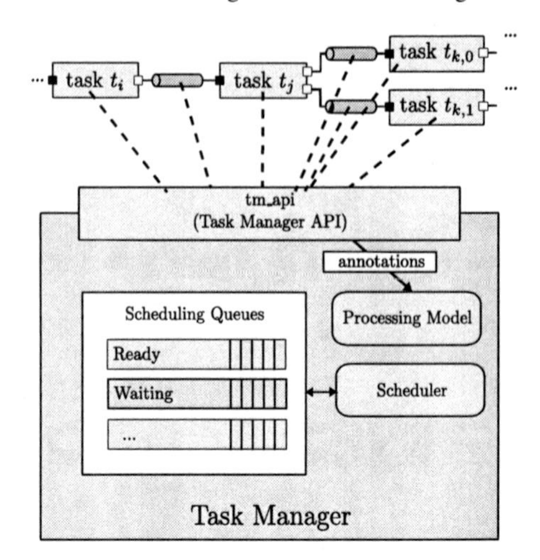

Fig. C.3 Stand-alone execution of task graph with the task manager (based on [307])

Instead of executing the task graph directly on the native SystemC scheduler, a global task manager is instantiated that serves as an intermediate scheduling layer. This manager mimics the execution of the task graph as if executed on a single, shared processing resource (Fig. C.3).

The integrated analysis framework features dynamic enabling and disabling of extensive tracing facilities to measure and analyze the system performance. One example is tracing the task execution over time. The obtained result plots visualize the execution order of tasks and gives valuable inside to the parallelism contained within the application executed. Figure C.4 illustrates such a trace.

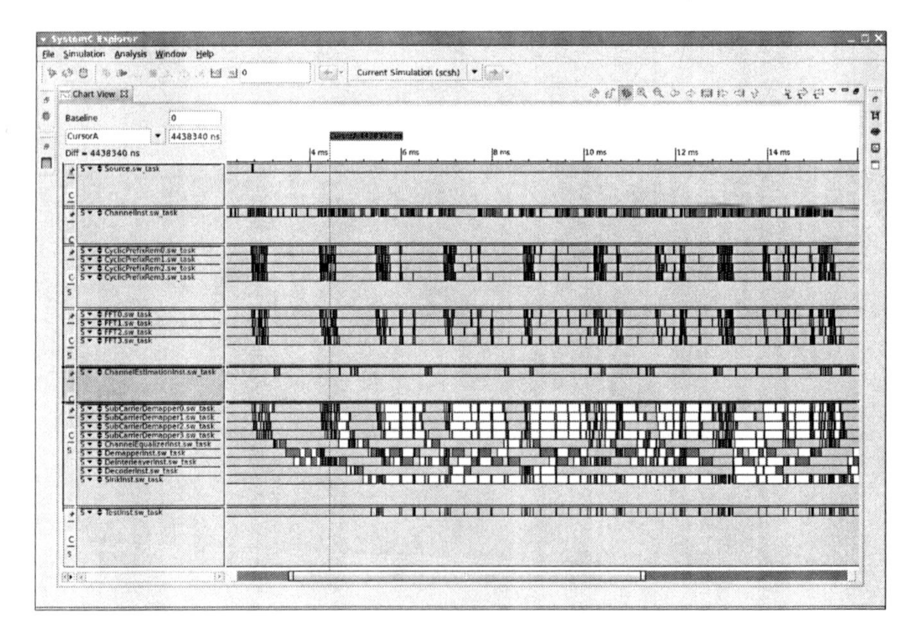

Fig. C.4 Task execution trace

C.3 VPU IP Component and Platform Modeling

At all times the modeled task graph can be mapped onto a hardware platform based on the VPU. The VPU is integrated into Platform Architect as a regular IP component. Apart from the component itself, the VPU library comprises a set of communication drivers, task schedulers, and different processing models for extended annotation techniques.

The VPU component is similar to a hierarchical SystemC model and its overall structure is illustrated in Fig. C.5. First of all, the VPU represents a shared processing resource, that is capable of executing a set of tasks concurrently. The execution order of tasks is controlled by the task manager and the scheduler component. The most important scheduling algorithms, like round-robin and priority scheduling, are supported. However, developers can easily overload and hence customize the scheduling algorithm. In addition, the VPU can be equipped with an arbitrary number of external communication ports compliant to the TLM standard 2.0 (TLM-2.0) [153]. These ports enable the modeling of external communication accesses that are caused by intertask communication and the task execution itself. The task execution and the implicitly occurring communication accesses during execution are captured by the processing model. Whenever an annotation is passed to the processing model, it adds a delay for the execution time and generates communication accesses to the external environment. For example, when emulating a processor core instruction fetches are generated over the instruction port. For tasks running on different processor cores that communicate with each other, the external communication accesses are modeled by drivers that explicitly read and write data from external memories and devices. Last but not least, the VPU model supports interrupt handling and possesses an external clock port for converting the annotations from discrete cycles to the time domain.

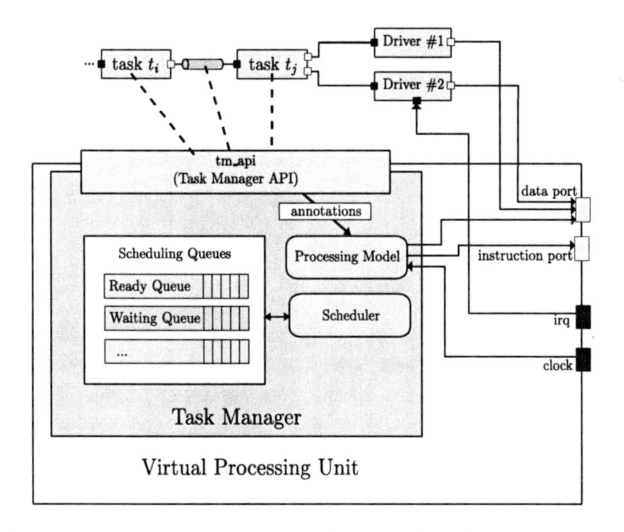

Fig. C.5 VPU IP component and task graph mapping (based on [307])

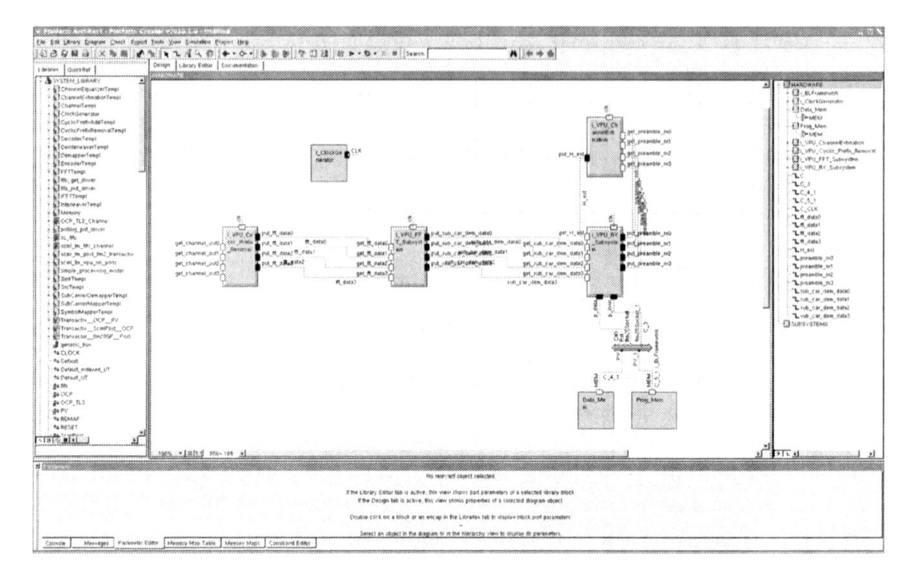

Fig. C.6 Hardware platform

In contrast to the time intensive modeling on a fine-grained level of abstraction, the VPU enables an abstract modeling that offers increased productivity by a simple and quick evaluation of various hardware options. The platform is composed out of different processing elements, the communication architecture, and the memory subsystem. While the latter two are typically modeled in detail, the processing elements executing the application are captured by VPUs. The modeling of the platform based on VPUs does not differ from regular platform development in Synopsys' PCT and the VPU IP component can be used arbitrarily along with other IP components. Figure C.6 illustrates an example platform incorporating multiple instances of the VPU component.

For investigation of the execution characteristic of an application running on the platform, the next step is to map the task graph onto the platform model that is discussed in the following section.

C.4 Task Graph Mapping

In a classical design environment with simulation techniques operating on instruction set simulation or even lower abstraction levels, the modeling of the hardware platform and the porting of the application is a time- and cost-intensive task. When using the VPU technology this step is significantly shortened, so that system architects can directly start performance investigations when the task graph and the platform model based on VPUs is available.

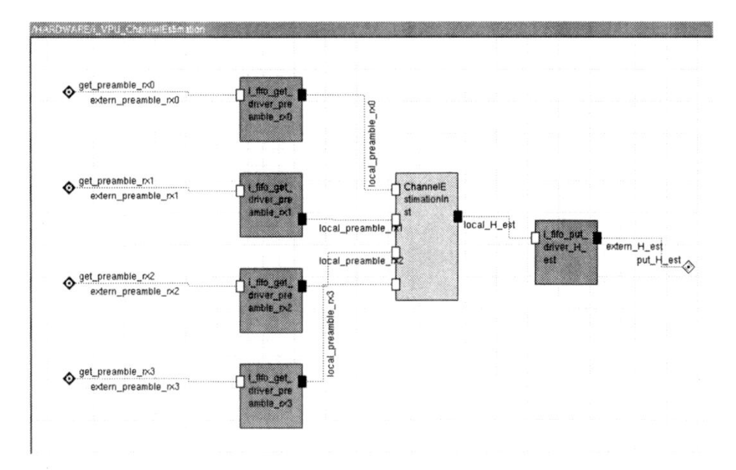

Fig. C.7 Mapped task graph (channel estimation subsystem)

Mapping the application, respectively the task graph, onto the hardware platform follows the basic principle as discussed in Sect. 7.5.2. Each task can be mapped to a processing element modeled as VPU by a simple drag and drop mechanism. When two tasks, which communicate over a common communication channel, are mapped to the same processing element (VPU) the communication can be handled internally within the component. Given the assumption that the underlying processing element is a processor core including a local memory, both tasks could for example share a common variable or FIFO implemented in software.

The more complicated case occurs when two tasks connected to one communication channel are distributed among different processing elements (VPUs) and data exchange must be accomplished over the external interfaces. Therefore, such connections have to be resolved by inserting drivers for sending and receiving data. These drivers can range from simple drivers accessing a shared memory to complex DMA-based driver implementations. Figure C.7 highlights a small subsystem containing one task with four inputs and one output. At the inputs, drivers for reading from hardware FIFOs are attached. Similarly, at the task output a driver is connected to write data to a hardware FIFO.

Containing both, the application and the hardware platform, the complete system is analyzed and the system architect can verify whether the performance requirements are kept.

References

1. Informa Telecoms & Media. http://www.informatm.com/.
2. M. Woh, Y. Lin, S. Seo, S. Mahlke, T. Mudge, C. Chakrabarti, R. Bruce, D. Kershaw, A. Reid, M. Wilder, and K. Flautner. From SODA to Scotch: The evolution of a wireless baseband processor. In *MICRO '08: Proceedings of the 2008 41st IEEE/ACM International Symposium on Microarchitecture*, pages 152–163, Washington, DC, USA, 2008. IEEE Computer Society.
3. G.E. Moore. Progress in digital electronics. *Technical Digest of the International Electron Devices Meeting*, IEEE Press, New York, 1975.
4. R.H. Dennard, F.H. Gaensslen, V.L. Rideout, E. Bassous, and A.R. LeBlanc. Design of ion-implanted MOSFET's with very small physical dimensions. *IEEE Journal of Solid-State Circuits*, 9(5):256–268, Oct. 1974.
5. G. Smith. Crisis of complexity. In *Gartner Dataquest briefing, 40th Design Automation Conference (DAC)*, June 2003.
6. J. Mitola, III. Cognitive radio for flexible mobile multimedia communications. *Mobile Networks and Applications*, 6(5):435–441, 2001.
7. A. Jerraya and W. Wolf. *Multiprocessor Systems-on-Chips (The Morgan Kaufmann Series in Systems on Silicon)*. Morgan Kaufmann, Los Altos, CA, Sept. 2004.
8. International Telecommunication Union (ITU). *http://www.itu.int/*, Jan. 2011.
9. European Telecommunications Standards Institute (ETSI). *http://www.etsi.org/*, Jan. 2011.
10. IEEE. *http://www.etsi.org/*, Jan. 2011.
11. H. Zimmermann. OSI reference model – the ISO model of architecture for open systems interconnection. *COM*, 28(4), April 1980.
12. Wireless Medium Access Control (MAC) and Physical Layer (PHY) Specifications for Wireless Personal Area Networks (WPANs). *IEEE Std. 802.15.1-2002*, 2002.
13. IEEE Standard for Information technology-Telecommunications and information exchange between systems-Local and metropolitan area networks-Specific requirements – Part 11: Wireless LAN Medium Access Control (MAC) and Physical Layer (PHY) Specifications. *IEEE Std 802.11-2007 (Revision of IEEE Std 802.11-1999)*, pages C1–C1184, Dec. 2007.
14. Apple Inc. *http://www.apple.com/*, Dec. 2007.
15. P. Mannion. Under the Hood: Inside the Apple iPhone. *EE Times, http://www.eetimes.com/news/latest/showArticle.jhtml?articleID=200001811*, Jan. 2007.
16. U. Ramacher. Software-defined radio prospects for multistandard mobile phones. *Computer*, 40(10):62–69, 2007.
17. LAN MAN Standards Commitee of the IEEE Computer Society. *IEEE Std 802.11a-1999*, Part 11: Wireless LAN Medium Access Control (MAC) and Physical Layer (PHY) Specifications: High Speed Physical Layer in the 5 GHz Band. 1999.
18. CoWare. SPD WLAN Library Reference (SPD2006.1), Feb. 2007.
19. G. Kahn. The semantics of a simple language for parallel programming. In J.L. Rosenfeld, editor, *Information Processing '74: Proceedings of the IFIP Congress*, pages 471–475. North-Holland, New York, NY, 1974.

T. Kempf et al., *Multiprocessor Systems on Chip: Design Space Exploration*,
DOI 10.1007/978-1-4419-8153-0,

20. E.A. Lee and D.G. Messerschmitt. Synchronous data flow. *Proceedings of the IEEE*, 75(9):1235–1245, 1987.
21. Z. Chamski. Parallelism and physical time constraints in multimedia applications, or another view on iterators (and arrays). In *Synchron'03*, Luminy, Dec. 2003.
22. J.D. Owens, D. Luebke, N. Govindaraju, M. Harris, J. Krueger, A.E. Lefohn, and T.J. Purcell. A survey of general-purpose computation on graphics hardware. *Computer Graphics Forum*, 26(1):80–113, 2007.
23. T. Wiegand, G.J. Sullivan, G. Bjontegaard, and A. Luthra. Overview of the H.264/AVC video coding standard. *IEEE Transactions on Circuits and Systems for Video Technology*, 13(7):560–576, July 2003.
24. J.L. Hennessy and D.A. Patterson. *Computer Architecture: A Quantitative Approach*. Morgan Kaufmann Publishers, Los Altos, CA, 4th ed. 2007.
25. C. Hammerschmidt. Intel starts foray into SoC market. Technical report, July 2008.
26. R. Weinreich and J. Sametinger. *Component-Based Software Engineering: Putting the Pieces Together*. Addison-Wesley Longman Publishing Co., Inc., Reading, MA, 2001.
27. Unified Modeling Language (UML). *http://www.uml.org*, Jan. 2011.
28. T. Kogel and H. Meyr. Heterogeneous MP-SoC – The solution to energy-efficient signal processing. In *Design Automation Conference (DAC)*, San Diego, USA, June 2004.
29. J.M. Rabaey. Wireless beyond the third generation-facing the energy challenge. In *International Symposium on Low Power Electronics and Design*, pages 1–3, 2001.
30. M. Gries and K. Keutzer. *Building ASIPs: The Mescal Methodology*. Springer, Berlin, Heidelberg, 2005.
31. E.M. Witte, T. Kempf, V. Ramakrishnan, and G. Ascheid, RWTH Aachen University, Germany; M. Adrat and M. Antweiler, Department of FKIE/KOM, Wachtberg, Germany. SDR Baseband Processing Portability: A Case Study. In *5th Karlsruhe Workshop on Software Radios (WSR'08)*, Karlsruhe, Germany, March 2008.
32. W.O. Cesario, D. Lyonnard, G. Nicolescu, Y. Paviot, S. Yoo, A.A. Jerraya, L. Gauthier, and M. Diaz-Nava. Multiprocessor SoC platforms: A component-based design approach. *IEEE Design and Test of Computers*, 19(6):52–63, Nov./Dec. 2002.
33. A. Sangiovanni-Vincentelli. Defining platform-based design. *EEDesign of EETimes*, Feb. 2002.
34. ARM. AMBA System Architecture. *http://www.arm.com/*, Jan. 2011.
35. IBM CoreConnect bus cores. *http://www.ibm.com/*, Jan. 2011.
36. Arteris Unveils Strategy, Technology for enabling Network on Chip (NoC) Design. Press Release, March 2003.
37. Texas Instruments. TI OMAP. *http://focus.ti.com/docs/prod/folders/print/omap3530.html*, Jan. 2011.
38. Texas Instruments Inc. OMAP 4430 Platform. *http://focus.ti.com/en/graphics/wtbu/OMAP4430-tn.gif*, Jan. 2011.
39. J. Kunkel. MPSoC IP integration and interoperability challenges. In *8th International Forum on Application-Specific Multi-Processor SoC*, June 2008.
40. J.A. de Oliveira and H. van Antwerpen. The Philips Nexperia digital video platform. In G. Martin and H. Chang, editors, *Winning the SoC Revolution: Experiences in Real Design*, Kluwer Academic Publishers, Boston, 2003.
41. Imagination Technologies Ltd. *POWERVR Graphics IP, http://www.imgtec.com/*, Jan. 2011.
42. B. Bailey, G. Martin, and A. Piziali. *ESL Design and Verification*. Morgan Kaufmann, Los Altos, CA, 1st ed., 2007.
43. J. Tourley. Survey says: software tools more important than chips, April 2005.
44. S.W. Smith. *The Scientist and Engineer's Guide to Digital Signal Processing*. California Technical Publishing, San Diego, CA, USA, 1997.
45. J.A. Fisher. Very long instruction word architectures and the ELI-512. In *ISCA '83: Proceedings of the 10th Annual International Symposium on Computer Architecture*, pages 140–150, Los Alamitos, CA, USA, 1983. IEEE Computer Society Press.
46. M. Flynn. Very high-speed computing systems. *Proceedings of the IEEE*, 54:1901–1909, Dec. 1966.

47. W.M. Johnson. *Superscalar Microprocessors Design*. Prentice Hall PTR, Englewood, Cliffs, NJ, 1990.
48. Tensilica. *http://www.tensilica.com/*, 2002.
49. A. Hofmann, H. Meyr, and R. Leupers. *Architecture Exploration for Embedded Processors with LISA*. PhD thesis, RWTH Aachen, 2002. ISBN 1-4020-7338-0.
50. A. Halambi, P. Grun, V. Ganesh, A. Khare, N. Dutt, and A. Nicolau. EXPRESSION: A language for architecture exploration through compiler/simulator retargetability. In *Proceedings of the Design Automation and Test in Europe Conference and Exhibition 1999*, pages 485–490, 1999.
51. M. Hohenauer, H. Scharwaechter, K. Karuri, O. Wahlen, T. Kogel, R. Leupers, G. Ascheid, H. Meyr, G. Braun, and H. van Someren. A methodology and tool suite for C compiler generation from ADL processor models. In *Proceedings of the Conference on Design, Automation and Test in Europe (DATE)*, Paris, France, Feb. 2004.
52. A. Wang, E. Killian, D. Maydan, and C. Rowen. Hardware/software instruction set configurability for system-on-chip processors. In *DAC '01: Proceedings of the 38th conference on Design automation*, pages 184–188, New York, NY, USA, 2001. ACM.
53. S. Hauck and A. Dehon, editors. Reconfigurable computing: The theory and practice of FPGA-based computation. *Systems on Silicon*. Morgan Kaufmann, Los Altos, CA, Nov. 2007.
54. Stretch Inc. *http://www.stretchinc.com/*, Jan. 2011.
55. A. Chattopadhyay, R. Leupers, H. Meyr, and G. Ascheid. *Language-driven Exploration and Implementation of Partially Re-configurable ASIPs*. Springer Publishing Company, Incorporated, Berlin (Heidelberg/New York), 2008.
56. R. Leupers, K. Karuri, S. Kraemer, and M. Pandey. A design flow for configurable embedded processors based on optimized instruction set extension synthesis. In *Proceedings of the International Conference on Design, Automation and Test in Europe (DATE)*, Munich, Germany, March 2006.
57. IEEE Standard VHDL Language Reference Manual. *IEEE Std 1076*, March 1987.
58. IEEE standard Verilog hardware description language. *IEEE Std. 1364-2001*, 2001.
59. T. Vogt and N. Wehn. A reconfigurable ASIP for convolutional and turbo decoding in an SDR environment. *IEEE Transactions on Very Large Scale Integration (VLSI) Systems*, 16(10):1309–1320, Oct. 2008.
60. T. Kempf, E.M. Witte, V. Ramakrishnan, G. Ascheid, M. Adrat, and M. Antweiler. A practical view of SDR baseband processing portability. In *Software Defined Radio Technical Conference (SDR'08)*, Washington, USA, Oct. 2008.
61. DSP-C Website. *http://www.dsp-c.org/*, Jan. 2011.
62. DSP-C Specification. *http://www.open-std.org/JTC1/SC22/WG14/www/docs/n854.pdf*, Oct. 1998.
63. ISO/IEC Working Group JTC1/SC22/WG14. Programming languages – C-Extensions to support embedded processors. Sep. 2007.
64. R. Leupers. C compiler retargeting. In Ienne, P. and Leupers, R. editors, *Customizable Embedded Processors: Design Technologies and Applications*. Morgan Kaufmann, Los Altos, CA, July 2006. Series in Systems on Silicon, ISBN 0-1236-9526-0.
65. A. Hemani, A. Jantsch, S. Kumar, A. Postula, J. Öberg, M. Millberg, D. Lindqvist. Network on a chip: An architecture for billion transistor era. In *Norchip Conference*, pages 166–173, November 2000.
66. OCP IP. *http://www.ocpip.org/*, Jan. 2011.
67. E. Salminen, A. Kulmala, and T.D. Hamalainen. *Survey of Network-on-chip Proposals*. White paper, OCP-IP, April 2008. Available online (13 pages).
68. J.D. Owens, W.J. Dally, R. Ho, D.N. Jayasimha, S.W. Keckler, and L.-S. Peh. Research challenges for on-chip interconnection networks. *IEEE Micro*, pages 96–108, Sep.–Oct. 2007.
69. A. Jantsch and H. Tenhunen, editors. *Networks on chip*. Kluwer Academic Publishers, Hingham, MA, USA, 2003.
70. P. Grun, A. Nicolau, and N. Dutt. *Memory Architecture Exploration for Programmable Embedded Systems*. Kluwer Academic Publishers, Norwell, MA, USA, 2002.

71. T. Vogt and N. Wehn. A reconfigurable application specific instruction set processor for convolutional and turbo decoding in a SDR environment. In *Proceedings of the International Conference on Design, Automation and Test in Europe (DATE)*, pages 38–43, New York, NY, USA, 2008. ACM.
72. BDTI Inc. Evaluating the DSP Capabilities of the Cortex-R4. *Inside DSP*, 2007.
73. M. Speth, H. Dawid, and F. Gersemsky. Design and verification challenges for 3G/3.5G/4G wireless baseband MPSoCs. In *MPSoC'08*, June 2008.
74. K. Deb. *Multi-Objective Optimization using Evolutionary Algorithms*. Wiley-Interscience Series in Systems and Optimization. Wiley, Chichester, 2001.
75. C.M. Christensen. *The innovator's dilemma: When new technologies cause great firms to fail.* Harvard Business School Press, Boston, MA, USA, 1997.
76. B. Kienhuis, E. Deprettere, K. Vissers, and P. van der Wolf. An approach for quantitative analysis of application-specific dataflow architectures. In *Proceedings of the IEEE Conference on Application Specific Architectures and Processors*, 1997.
77. M. Gries. Methods for Evaluating and Covering the Design Space during Early Design Development. Technical Report UCB/ERL M03/32, Electronics Research Lab, University of California at Berkeley, Aug. 2003.
78. IEEE standard computer dictionary. A compilation of IEEE standard computer glossaries. *IEEE Std 610*, pages –, Jan. 1991.
79. A. Jantsch. *Modeling Embedded Systems and SoC's: Concurrency and Time in Models of Computation*. Morgan Kaufmann Publishers Inc., San Francisco, CA, USA, 2003.
80. S. Edwards, L. Lavagno, E.A. Lee, and A. Sangiovanni-vincentelli. Design of embedded systems: Formal models, validation, and synthesis. In *Proceedings of the IEEE*, pages 366–390, 1997.
81. E.A. Lee and A. Sangiovanni-Vincentelli. Comparing models of computation. In *ICCAD '96: Proceedings of the 1996 IEEE/ACM International Conference on Computer-aided Design*, Washington, DC, USA, 1996. IEEE Computer Society.
82. G.S. Fishman. *Principles of Discrete Event Simulation*. Wiley, New York, NY, USA, 1978.
83. E.A. Lee and T.M. Parks. *Dataflow Process Networks*. pages 59–85, 2002.
84. A.L. Davis and R.M. Keller. Data flow program graphs. *Computer*, 15(2):26–41, Feb. 1982.
85. E.A. Lee and D.G. Messerschmitt. Static scheduling of synchronous data flow programs for digital signal processing. *IEEE Transactions on Computers*, 36(1):24–35, 1987.
86. S. Sriram and S.S. Bhattacharyya. *Embedded Multiprocessors: Scheduling and Synchronization*. Marcel Dekker, Inc., New York, NY, USA, 2000.
87. J.T. Buck. *Scheduling Dynamic Dataflow Graphs with Bounded Memory using the Token Flow Model*. PhD thesis, EECS UC Berkeley, 1993.
88. T. Grötker, R. Schoenen, and H. Meyr. *PCC: A Modeling Technique for Mixed Control/Data Flow Systems*. pages 482–486, Mar. 1997.
89. R. Milner. *A Calculus of Communicating Systems*. Springer-Verlag New York, Inc., Secaucus, NJ, USA, 1982.
90. C.A.R. Hoare. Communicating sequential processes. *Communications of the ACM*, 21(8):666–677, 1978.
91. Specification and Description Language, 1987. ITU-T Recommendation Z. 100.
92. P. Le Guernic, A. Benveniste, P. Bournai, and T. Gautier. Signal – A data flow-oriented language for signal processing. *IEEE Transactions on Acoustics, Speech and Signal Processing*, 34(2):362–374, April 1986.
93. N. Halbwachs, P. Caspi, P. Raymond, and D. Pilaud. The synchronous data flow programming language LUSTRE. *Proceedings of the IEEE*, 79(9):1305–1320, Sept. 1991.
94. G. Berry. *The Foundations of Esterel*. pages 425–454, MIT Press, Cambridge, MA, 2000.
95. F. Maraninchi. The Argos language: graphical representation of automata and description of reactive systems. In *IEEE Workshop on Visual Languages*, Kobe, Japan, Oct. 1991.
96. R. Lipsett, C.A. Ussery, and C.F. Schaefer. *VHDL, Hardware Description and Design*. Kluwer Academic Publishers, Norwell, MA, USA, 1993.
97. T. Grötker, S. Liao, G. Martin, S. Swan. *System Design with SystemC*. Kluwer Academic Publishers, Dordrecht (Hingham, MA), 2002.

98. W. Wolf. A decade of hardware/software codesign. *IEEE Computer*, 36(4):38–43, April 2003.
99. S.A. Edwards, L. Lavagno, E.A. Lee, and A. Sangiovanni-Vincentelli. Design of embedded systems: Formal models, validation, and synthesis. *Proceedings of the IEEE*, 85(3):366–390, March 1997.
100. T. Kogel, R. Leupers, and H. Meyr. *Integrated System-Level Modeling of Network-on-Chip enabled Multi-Processor Platforms*. Springer-Verlag New York, Inc., Secaucus, NJ, USA, 2006.
101. J.A. Rowson. Hardware/Software Co-Simulation. In *Proceedings of the Design Automation Conference (DAC)*, 1994.
102. J. Buck, S. Ha, E.A. Lee, and D.G. Messerschmitt. *Ptolemy: A Framework for Simulating and Prototyping Heterogeneous Systems*. Kluwer Academic Publishers, Norwell, MA, USA, 2002.
103. F. Balarin, P.D. Giusto, A. Jurecska, C. Passerone, E. Sentovich, B. Tabbara, M. Chiodo, H. Hsieh, L. Lavagno, A. Sangiovanni-Vincentelli, and K. Suzuki. *Hardware-Software Co-Design of Embedded Systems: The POLIS Approach*. Springer-Verlag Gmbh, Berlin Heidelberg, 1997.
104. G. Martin and J.-Y. Brunel. Platform-based co-design and co-development: Experience methodology and trends. In *Electronic Design Process Workshop*, Monterey, CA, USA, 2002.
105. Synopsys System Studio. *http://www.synopsys.com/*, Jan. 2011.
106. Mentor Graphics Seamless. *http://www.mentor.com/*, Jan. 2011.
107. W. Mueller, J. Ruf, D. Hoffmann, J. Gerlach, T. Kropf, and W. Rosenstiehl. The simulation semantics of SystemC. In *Proceedings of the International Conference on Design, Automation and Test in Europe (DATE)*, 2001.
108. A. Müller, T. Kogel, and G. Post. Methodology for ATM-Cell processing system design. In *12th Annual 1999 IEEE International ASIC/SOC Conference*, Washington, DC, Sept. 1999.
109. R.K. Gupta, C.N. Coelho, Jr., and G. De Micheli. Synthesis and simulation of digital systems containing interacting hardware and software components. In *DAC '92: Proceedings of the 29th ACM/IEEE conference on Design automation*, pages 225–230, Los Alamitos, CA, USA, 1992. IEEE Computer Society Press.
110. R. Ernst, J. Henkel, and T. Benner. Hardware-software cosynthesis for microcontrollers. *IEEE Design and Test of Computers*, 10(4):64–75, Dec. 1993.
111. J. Madsen, J. Grode, P.V. Knudsen, M.E. Petersen, and A. Haxthausen. LYCOS: The Lyngby co-synthesis system. *Design Automation of Embedded Systems*, 2(2):195–236, 1997.
112. T.B. Ismail, M. Abid, and A. Jerraya. COSMOS: A codesign approach for communicating systems. In *Third International Workshop on Hardware/Software Codesign*, pages 17–24, Silver Spring, MD, Sept. 1994. IEEE Computer Society Press.
113. T.-Y. Yen and W. Wolf. Communication synthesis for distributed embedded systems. In *IC-CAD '95: Proceedings of the 1995 IEEE/ACM International Conference on Computer-aided Design*, pages 288–294, Washington, DC, USA, 1995. IEEE Computer Society.
114. K. Lahiri, A. Raghunathan, and S. Dey. Fast performance analysis of bus-based system-on-chip communication architectures. In *Proceedings of the IEEE International Conference on Computer Aided Design*, 1999.
115. D. Bertozzi, A. Jalabert, S. Murali, R. Tamhankar, S. Stergiou, L. Benini, and G. De Micheli. Noc synthesis flow for customized domain specific multiprocessor systems-on-chip. *IEEE Transactions on Parallel and Distributed Systems*, 16(2):113–129, Feb. 2005.
116. D. Bertozzi and L. Benini. Xpipes: A network-on-chip architecture for gigascale systems-on-chip. *IEEE Circuits and Systems Magazine*, 4(2):18–31, 2004.
117. S. Murali and G. De Micheli. SUNMAP: A tool for automatic topology selection and generation for NoCs. In *DAC '04: Proceedings of the 41st Annual Conference on Design Automation*, pages 914–919, New York, NY, USA, 2004. ACM.
118. K. Van Rompaey, D. Verkest, I. Bolsens, and H. De Man. CoWare – A design environment for heterogeneous hardware/software systems. In *Proceedings of the European Design Automation Conference (EuroDAC)*, 1996.

119. F. Balarin, Y. Watanabe, H. Hsieh, L. Lavagno, C. Passerone, and A. Sangiovanni-Vincentelli. Metropolis: An integrated electronic system design environment. *IEEE Computer*, 36(4): 45–52, April 2003.
120. F. Balarin, L. Lavagno, C. Passerone, and Y. Watanabe. Processes, interfaces and platforms. Embedded software modeling in Metropolis. In *Proceedings of EMSOFT'02*, October 2002.
121. P. Lieverse, P. van der Wolf, E. Deprettere, and K. Vissers. A methodology for architecture exploration of heterogeneous signal processing systems. In *Proceedings of IEEE Workshop on Signal Processing Systems SiPS 99*, pages 181–190, 1999.
122. E. Deprettere P. Lieverse, P. van der Wolf, and K. Vissers. A methodology for architecture exploration of heterogeneous signal processing systems. *Journal of VLSI Signal Processing for Signal, Image and Video Technology*, 29(3):197–207, Nov. 2001.
123. E.A. de Kock, W.J.M. Smits, P. van der Wolf, J.-Y. Brunel, W.M. Kruijtzer, P. Lieverse, K.A. Vissers, and G. Essink. YAPI: Application modeling for signal processing systems. In *Proceedings of the Design Automation Conference (DAC)*, pages 402–405. ACM Press, New York, 2000.
124. A.D. Pimentel, L.O. Hertzberger, P. Lieverse, P. van der Wolf, and E.F. Deprettere. Exploring embedded-systems architectures with artemis. *IEEE Computer*, 34(11):57–63, Nov. 2001.
125. S. Polstra. A systematic approach to exploring embedded system architectures at multiple abstraction levels. *IEEE Transactions on Computers*, 55(2):99–112, 2006. Andy D. Member-Pimentel and Cagkan Erbas.
126. A.D. Pimentel and C. Erbas. An IDF based trace transformation method for communication refinement. In *Proceedings of the Design Automation Conference (DAC)*, June 2003.
127. C. Erbas, S.C. Erbas, and A.D. Pimentel. A multiobjective optimization model for exploring multiprocessor mappings of process networks. In *Proceedings of the IEEE/ACM/IFIP International Conference on Hardware/Software Codesign and System Synthesis*, Oct. 2003.
128. V.D. Zivkovic, E. Deprettere, P. van der Wolf, and E. de Kock. Design space exploration of streaming multiprocessor architectures. In *Proceedings of the IEEE Workshop on Signal Processing Systems (SIPS '02)*, pages 228–234, Oct. 2002.
129. H. Nikolov, M. Thompson, T. Stefanov, A. Pimentel, S. Polstra, R. Bose, C. Zissulescu, and E. Deprettere. Daedalus: Toward composable multimedia mp-soc design. In *DAC '08: Proceedings of the 45th Annual Conference on Design Automation*, pages 574–579, New York, NY, USA, 2008. ACM.
130. M.J. Rutten, J.T.J. van Eijndhoven, E.G.T. Jaspers, P. van der Wolf, O.P. Gangwal, A. Timmer, and E.-J.D. Pol. A heterogeneous multiprocessor architecture for flexible media processing. *IEEE Design and Test of Computers*, 19(4):39–50, July/Aug. 2002.
131. J.M. Paul and D.E. Thomas. A layered, codesign virtual machine approach to modeling computer systems. In *Proceedings of the International Conference on Design, Automation and Test in Europe (DATE)*, 2002.
132. A.S. Cassidy, J.M. Paul, and D.E. Thomas. Layered, multi-threaded, high-level performance design. In *Proceedings of the International Conference on Design, Automation and Test in Europe (DATE)*, 2003.
133. J.M. Paul, A. Bobrek, J.E. Nelson, J.J. Pieper, and D.E. Thomas. Schedulers as model-based design elements in programmable heterogeneous multiprocessors. In *Proceedings of the Design Automation Conference (DAC)*, 2003.
134. S. Mahadevan, M. Storgaard, J. Madsen, and K. Virk. ARTS: A system-level framework for modeling MPSoC components and analysis of their causality. In *Proceedings of MASCOTS'05*, pages 480–483, Sept. 2005.
135. J. Madsen, S. Mahadevan, K. Virk, and M. Gonzalez. Network-on-chip modeling for system-level multiprocessor simulation. In *RTSS '03: Proceedings of the 24th IEEE International Real-Time Systems Symposium*, page 265, Washington, DC, USA, 2003. IEEE Computer Society.
136. J. Madsen, K. Virk, and S. Mahadevan. Abstract system-on-chip modelling in systemc. In *European SystemC Users Group Meeting (DATE 2004)*, April 2004.

137. K. Virk, J. Madsen, and M.J. Gonzalez. Abstract RTOS modelling for multiprocessor system-on-chip. In *International Symposium on System-on-Chip*, pages 147–150, New York. IEEE, Nov. 2003.
138. E. Bensoudane, P.G. Paulin, and C. Pilkington. StepNP: A system-level exploration platform for network processors. *IEEE Design and Test of Computers*, 19(6):17–26, Nov.–Dec. 2002.
139. R. Morris, E. Kohler, J. Jannotti, and M.F. Kaashoek. The Click modular router. *SIGOPS Operating Systems Review*, 33(5):217–231, 1999.
140. K. Keutzer N. Shah, and W. Plishker. NP-Click: A programming model for the Intel IXP1200. In *2nd Workshop on Network Processors (NP-2) at the 9th International Symposium on High Performance Computer Architecture (HPCA-9), Anaheim, CA*, Feb. 2003.
141. F.R. Wagner, W. Cesário, and A.A. Jerraya. Hardware/software IP integration using the ROSES design environment. *Transaction on Embedded Computing System*, 6(3):17, 2007.
142. R. Dömer. *System-level Modeling and Design with the SpecC Language*. PhD thesis, University Dortmund, 2000.
143. Open SystemC Initiative (OSCI), http://www.systemc.org, Jan. 2011.
144. SpecC Technology Consortium. *http://www.specc.org, 2002.*
145. Standard for SystemVerilog - Unified Hardware Design, Specification, and Verification Language. *IEC 62530:2007 (E)*, pages 1–668, 2007.
146. P.L. Flake and S.J. Davidmann. Superlog, a unified design language for system-on-chip. In *Proceedings of the Asia South Pacific Design Automation Conference (ASPDAC)*, pages 583–586, New York, NY, USA, 2000. ACM.
147. W. Müller, W. Rosenstiel, and J. Ruf, editors, *SystemC – Methodologies and Applications*, Kluwer Academic Publishers, Dordrecht, June 2003.
148. K. Keutzer, S. Malik, A.R. Newton, J.M. Rabaey, and A. Sangiovanni-Vincentelli. System-level design: Orthogonalization of concerns and platform-based design. *IEEE Transactions on Computer-Aided Design of Integrated Circuits and Systems*, 19(12):1523–1543, Dec. 2000.
149. D.C. Black and J. Donovan. *SystemC from the Ground up*. Kluwer Academic Publishers, Dordrecht, 2004.
150. M. Birnbaum and H. Sachs. How VSIA answers the SOC dilemma. *IEEE Computer*, 32(6):42–50, Jun 1999.
151. T. Kogel, A. Haverinen, and J. Aldis. OCP TLM for Architectural Modeling, OCP-IP, http://www.ocpip.org/. Technical report, 2005.
152. Open Core Protocol International Partnership (OCP-IP). *OCP datasheet, http://www.ocpip.org*, Jan. 2011.
153. Open SystemC Initiative (OSCI). Transaction Level Modeling (TLM) Library, Release 2.0, 2008.
154. E.M. Witte, T. Kempf, V. Ramakrishnan, and G. Ascheid, RWTH Aachen University, Germany; M. Adrat and M. Antweiler, Department of FKIE/KOM, Wachtberg, Germany. A seamless software defined radio development flow for waveform and prototype debugging. In *02/2008 Journal of Telecommunications and Information Technology (JTIT)*, Warsaw, Poland, 2008.
155. G.R. Hellestrand. The revolution in systems engineering. *IEEE Spectrum* , 36(9):43–51, Sept. 1999.
156. K. Keutzer, A.R. Newton, J.M. Rabaey, and A. Sangiovanni-Vincentelli. System-level design: Orthogonalization of concerns and platform-based design. *IEEE Journal of Computer Aided Design*, 19(12):1523–1543, 2000.
157. Synopsys DesignWare IP. http://www.synopsys.com, Jan. 2011.
158. CoWare Model Library. *http://www.coware.com/*, Jan. 2011.
159. Doulos. *http://www.doulos.com/*, Jan. 2011.
160. G. Braun, A. Nohl, A. Hoffmann, O. Schliebusch, R. Leupers, and H. Meyr. A universal technique for fast and flexible instruction-set architecture simulation. *IEEE Transactions on Computer-Aided Design of Integrated Circuits and Systems*, 23(12):1625–1639, Dec. 2004.
161. A. Wieferink, H. Meyr, and R. Leupers. *Retargetable Processor System Integration into Multi-Processor System-on-Chip Platforms*. Springer Publishing Company, Inc., Berlin, Heidelberg, 2008.

162. L. Gao, S. Kraemer, R. Leupers, G. Ascheid, and H. Meyr. A fast and generic hybrid simulation approach using C virtual machine. In *Proceedings of the Conference on Compilers, Architecture, and Synthesis for Embedded Systems (CASES '07)*, Salzburg, Austria, Oct. 2007.
163. M. Burtscher and I. Ganusov. Automatic synthesis of high-speed processor simulators. In *37th International Symposium on Microarchitecture, 2004. MICRO-37 2004.*, pages 55–66, Dec. 2004.
164. J. Zhu and D.D. Gajski. A retargetable, ultra-fast instruction set simulator. In *Proceedings of the International Conference on Design, Automation and Test in Europe (DATE)*, page 62, New York, NY, USA, 1999. ACM.
165. A. Nohl, G. Braun, A. Hoffmann, O. Schliebusch, R. Leupers, and H. Meyr. A universal technique for fast and flexible instruction-set architecture simulation. In *Proceedings of the Design Automation Conference (DAC)*, 2002.
166. B. Cmelik and D. Keppel. Shade: A fast instruction-set simulator for execution profiling. In *SIGMETRICS '94: Proceedings of the 1994 ACM SIGMETRICS Conference on Measurement and Modeling of Computer Systems*, pages 128–137, New York, NY, USA, 1994. ACM.
167. W. Qin, J. D'Errico, and X. Zhu. A multiprocessing approach to accelerate retargetable and portable dynamic-compiled instruction-set simulation. In *Proceedings of the IEEE/ACM/IFIP International Conference on Hardware/Software Codesign and System Synthesis*, pages 193–198, New York, NY, USA, 2006. ACM.
168. M. Reshadi, P. Mishra, and N. Dutt. Instruction set compiled simulation: A technique for fast and flexible instruction set simulation. In *DAC '03: Proceedings of the 40th Conference on Design Automation*, pages 758–763, New York, NY, USA, 2003. ACM.
169. N.P. Topham and D. Jones. High speed CPU simulation using JIT binary translation. In *3rd Annual Workshop on Modeling, Benchmarking and Simulation, held in Conjunction with ISCA-34*, San Diego CA, USA, June 2007.
170. K. Torsten, K. Kingshuk, and G. Lei. Software instrumentation. In B. Wah, editor, *Wiley Encyclopedia of Computer Science and Engineering* , Wiley, Hoboken, Dec. 2008.
171. CoWare Platform Architect. *http://www.coware.com/*, Jan. 2011.
172. The Eclipse Foundation. Eclipse IDE. *http://www.eclipse.org/*, Jan. 2011.
173. CoWare Platform Creator. *http://www.coware.com/*, Jan. 2011.
174. ARM Ltd. ARM Embedded Processors. *http://www.arm.com/*, Jan. 2011.
175. T. Kogel, M. Doerper, T. Kempf, A. Wieferink, R. Leupers, G. Ascheid, and H. Meyr. Virtual architecture mapping: A systemc based methodology for architectural exploration of system-on-chip designs. In *SAMOS*, pages 138–148, 2004.
176. Synopsys. Synopsys Innovator. *http://www.synopsys.com/*, Jan. 2011.
177. Carbon Design Systems Inc. *http://carbondesignsystems.com/*, Jan. 2011.
178. Virtutech Simics. *http://www.virtutech.com/*, Jan. 2011.
179. VaSt Systems. *http://www.vastsystems.com/*, Jan. 2011.
180. Triton Tuner. Poseidon, *http://www.poseidon-systems.com/*, Jan. 2011.
181. Open Virtual Platforms (OVP). *http://www.ovpworld.org/*, Jan. 2011.
182. GreenSocs. *http://www.greensocs.com/*, Jan. 2011.
183. M. Coppola, S. Curaba, M.D. Grammatikakis, G. Maruccia, and F. Papariello. The OCCN user manual. Technical report, Dec. 2003.
184. M. Coppola, S. Curaba, M.D. Grammatikakis, G. Maruccia, and F. Papariello. Occn: A network-on-chip modeling and simulation framework. In *Proceedings of Design, Automation and Test in Europe Conference and Exhibition, volume 3*, pages 174–179 Vol.3, 2004.
185. J.D. Ullman. NP-Complete Scheduling Problems. *Journal of Computer and System Sciences*, 10(3):384–393, 1975.
186. H. Kasahara and S. Narita. Practical Multiprocessor Scheduling Algorithms for Efficient Parallel Processing. *IEEE Trans. Comput.*, 33(11):1023–1029, 1984.
187. V. Chaudhary and J.K. Aggarwal. A generalized scheme for mapping parallel algorithms. *IEEE Transactions on Parallel and Distributed Systems*, 4(3):328–346, March 1993.
188. E.D. Lazowska, J. Zahorjan, G.S. Graham, and K.C. Sevcik. *Quantitative System Performance, Computer System Analysis Using Queuing Network Models: Computer Analysis Using Queuing Network Models*. Prentice Hall, Englewood, Cliffs, NJ, Feb. 1984.

189. F. Baccelli, G. Cohen, G.J. Olsder, and J.-P. Quadrat. *Synchronization and Linearity: An Algebra for Discrete Event Systems*. Wiley, London, 2nd ed., Oct. 2001.
190. P. Thiran, J.-Y. Le Boudec. *Network Calculus – A Theory of Deterministic Queuing Systems for the Internet*. Springer-Verlag GmbH, Berlin, Heidelberg, Feb. 2007.
191. L. Thiele, S. Chakraborty, and M. Naedele. Real-time calculus for scheduling hard real-time systems. In *Proceedings of the International Symposium on ISCAS 2000 Geneva, The 2000 IEEE Circuits and Systems* , volume 4, pages 101–104, 2000.
192. L. Thiele, S. Chakraborty, M. Gries, A. Maxiaguine, and J. Greutert. Embedded software in network processors – models and algorithms. In *Proceedings of the First Workshop on Embedded Software (EMSOFT)*, pages 416–434, Lake Tahoe, California, USA, Oct. 2001. Springer-Verlag.
193. L. Thiele, S. Chakraborty, M. Gries, and S. Künzli. A framework for evaluating design tradeoffs in packet processing architectures. In *39th Design Automation Conference (DAC 2002)*, pages 880–885, New Orleans LA, USA, June 2002. ACM.
194. L. Thiele, S. Chakraborty, M. Gries, and S. Künzli. Design space exploration of network processor architectures. In *Network Processor Design: Issues and Practices, Volume 1*, pages 55–89. 2002.
195. S. Chakraborty, S. Künzli, and L. Thiele. A general framework for analysing system properties in platform-based embedded system designs. In *Proceedings of the International Conference on Design, Automation and Test in Europe (DATE)*, pages 190–195, Munich, Germany, March 2003. IEEE.
196. K. Richter, D. Ziegenbein, M. Jersak, and R. Ernst. Bottom-up performance analysis of HW/SW platforms. In *Proceedings of the IFIP 17th World Computer Congress – TC10 Stream on Distributed and Parallel Embedded Systems DIPES '02*, Deventer, The Netherlands, 2002.
197. R. Henia, A. Hamann, M. Jersak, R. Racu, K. Richter, and R. Ernst. System level performance analysis – the symTA/s approach. *IEE Proceedings Computers and Digital Techniques*, 152(2):148–166, 2005.
198. K. Richter. *Compositional Scheduling Analysis Using Standard Event Models*. PhD thesis, Technical University of Braunschweig, 2004.
199. S. Künzli. *Efficient Design Space Exploration for Embedded Systems*. PhD thesis, ETH Zurich, April 2006.
200. S. Chakraborty. *System-Level Timing Analysis and Scheduling for Embedded Packet Processors*. PhD thesis, ETH Zurich, April 2003.
201. P. Ienne. Analytical models of communication for MPSoCs. In *MPSoC'08*, June 2008.
202. P. Pop, P. Eles, Z. Peng, and T. Pop. Analysis and optimization of distributed real-time embedded systems. *ACM Transactions on Design Automation of Electronic Systems*, July 2006.
203. P. Eles, K. Kuchcinski, Z. Peng, A. Doboli, and P. Pop. Scheduling of conditional process graphs for the synthesis of embedded systems. In *Proceedings of Design, Automation and Test in Europe*, pages 132–138, 1998.
204. P. Eles, A. Doboli, P. Pop, and Z. Peng. Scheduling with bus access optimization for distributed embedded systems. *IEEE Transactions on VLSI Systems*, 8(5):472–491, 2000.
205. M. Bekooij, S. Parmar, and J. van Meerbergen. Performance guarantees by simulation of process. In *SCOPES '05: Proceedings of the 2005 workshop on Software and compilers for embedded systems*, pages 10–19, New York, NY, USA, 2005. ACM.
206. R.A. Uhlig and T.N. Mudge. Trace-driven memory simulation. *ACM Computing Surveys*, 29(2):128–170, June 1997.
207. W. Fornaciari, D. Sciuto, C. Silvano, and V. Zaccaria. A design framework to efficiently explore energy-delay tradeoffs. In *Proceedings of the International Symposium on Hardware/Software Codesign (CODES)*, 2001.
208. T.D. Givargis, J. Henkel, and F. Vahid. Interface and cache power exploration for core-based embedded system design. In *IEEE/ACM International Conference on Proceedings of the Digest of Technical Papers Computer-Aided Design 1999*, pages 270–273, 1999.

209. M.A. Franklin and T. Wolf. A network processor performance and design model with benchmark parameterization. In *Proceedings of Network Processor Workshop in Conjunction with Eighth International Symposium on High Performance Computer Architecture (HPCA-8)*, pages 63–74, Cambridge, MA, Feb. 2002.
210. K. Lahiri, K. Lahiri, A. Raghunathan, and S. Dey. Performance analysis of systems with multi-channel communication architectures. In *Proceedings of the Thirteenth International Conference on VLSI Design*, 2000.
211. M. Ariyamparambath, D. Bussagila, B. Reinkemeier, T. Kogel, and T. Kempf. A highly efficient modeling style for heterogeneous bus architectures. In *International Symposium on System-on-Chip*, Tampere (Finland), Nov. 2003.
212. T. Kogel, M. Doerper, A. Wieferink, R. Leupers, G. Ascheid, H. Meyr, and S. Goossens. A modular simulation framework for architectural exploration of on-chip interconnection networks. In *The First IEEE/ACM/IFIP International Conference on HW/SW Codesign and System Synthesis*, Newport Beach (California USA), Oct. 2003.
213. V.D. Zivkovic, P. van der Wolf, E.F. Deprettere, and E.A. de Kock. Design space exploration of streaming multiprocessor architectures. In *Proceedings of IEEE International Workshop on Signal Processing Systems (SIPS)*, Oct. 2002.
214. A. Bobrek, J.J. Pieper, J.E. Nelson, J.M. Paul, and D.E. Thomas. Modeling shared resource contention using a hybrid simulation/analytical approach. In *Proceedings of the International Conference on Design, Automation and Test in Europe (DATE)*, page 21144, Washington, DC, USA, 2004. IEEE Computer Society.
215. T. Wolf and M.A. Franklin. Performance models for network processor design. *IEEE Transactions on Parallel and Distributed Systems*, 17(6):548–561, 2006.
216. M.A. Franklin and T. Wolf. Power considerations in network processor design. In *Proceedings of Network Processor Workshop in Conjunction with Ninth International Symposium on High Performance Computer Architecture (HPCA-9)*, pages 10–22, Anaheim, CA, Feb. 2003.
217. S. Künzli, F. Poletti, L. Benini, and L. Thiele. Combining simulation and formal methods for system-level performance analysis. In *Proceedings of the International Conference on Design, Automation and Test in Europe (DATE)*, pages 236–241, 3001 Leuven, Belgium, Belgium, 2006. European Design and Automation Association.
218. V. Pareto. Manuel déconomie politique. *Bullitan of American Mathematical Society*, 18: 462–474, 1912.
219. Synopsys Inc. *http://www.synopsys.com*, Jan. 2011.
220. Cadence Design Systems Inc. *http://www.cadence.com/*, Jan. 2011.
221. Magma Design Automation Inc. *http://www.magma-da.com/*, Jan. 2011.
222. J.A. Rowson and A. Sangiovanni-Vincentelli. Interface-Based Design. In *Proceedings of the Design Automation Conference (DAC)*, 1997.
223. B. Kienhuis, E. Deprettere, K. Vissers, and P. van der Wolf. An approach for quantitative analysis of application-specific dataflow architectures. In *Proceedings of the IEEE International Conference on Application-Specific Systems, Architectures and Processors*, pages 338–349, 1997.
224. W. Wolf. A decade of hardware/software codesign. *Computer*, 36(4):38–43, 2003.
225. S. Verdoolaege, H. Nikolov, and T. Stefanov. Pn: A tool for improved derivation of process networks. *EURASIP Journal of Embedded Systems*, 2007(1):19–19, 2007.
226. V. Reyes, T. Bautista, G. Marrero, P.P. Carballo, and W. Kruijtzer. CASSE: A system-level modeling and design-space exploration tool for multiprocessor systems-on-chip. In *Proceedings of DSD*, pages 476–483, Aug. 2004.
227. P. van der Wolf, E. de Kock, T. Henriksson, W. Kruijtzer, and G. Essink. Design and programming of embedded multiprocessors: An interface-centric approach. In *Procedings of the IEEE/ACM/IFIP International Conference on Hardware/Software Codesign and System Synthesis*, pages 206–217, New York, NY, USA, 2004. ACM.
228. V. Reyes, W. Kruijtzer, T. Bautista, G. Alkadi, and A. Nú nez. A unified system-level modeling and simulation environment for mpsoc design: Mpeg-4 decoder case study.

In *Proceedings of the International Conference on Design, Automation and Test in Europe (DATE)*, pages 474–479, 3001 Leuven, Belgium, Belgium, 2006. European Design and Automation Association.
229. W. Tibboel, V. Reyes, M. Klompstra, and D. Alders. System-level design flow based on a functional reference for hw and sw. In *Proceedings of the Design Automation Conference (DAC)*, pages 23–28, New York, NY, USA, 2007. ACM.
230. H. Nikolov, T. Stefanov, and E. Deprettere. Multi-processor system design with espam. In *Proceedings of the IEEE/ACM/IFIP International Conference on Hardware/Software Codesign and System Synthesis*, pages 211–216, New York, NY, USA, 2006. ACM.
231. T. Stefanov, C. Zissulescu, A. Turjan, B. Kienhuis, and E. Deprettere. System design using kahn process networks: The compaan/laura approach. In *Proceedings of the International Conference on Design, Automation and Test in Europe (DATE)*, page 10340, Washington, DC, USA, 2004. IEEE Computer Society.
232. K. Popovici, X. Guerin, F. Rousseau, P.S. Paolucci, and A.A. Jerraya. Platform-based software design flow for heterogeneous mpsoc. *Transactions on Embedded Computing Systems*, 7(4):1–23, 2008.
233. L. Gauthier, S. Yoo, and A.A. Jerraya. Automatic generation and targeting of application-specific operating systems and embedded systems software. *IEEE Transactions on Computer-Aided Design of Integrated Circuits and Systems*, 20(11):1293–1301, Nov. 2001.
234. G. Schirner and R. Dömer. Result-oriented modeling – a novel technique for fast and accurate tlm. *IEEE Transactions on CAD of Integrated Circuits and Systems*, 26(9):1688–1699, 2007.
235. G. Schirner, A. Gerstlauer, and R. Dömer. Automatic generation of hardware dependent software for mpsocs from abstract system specifications. In *Proceedings of the Asia South Pacific Design Automation Conference (ASPDAC)*, pages 271–276, Los Alamitos, CA, USA, 2008. IEEE Computer Society Press.
236. C. Haubelt, J. Falk, J. Keinert, T. Schlichter, M. Streubühr, A. Deyhle, A. Hadert, and J. Teich. A systemc-based design methodology for digital signal processing systems. *EURASIP Journal Embedded Systems*, 2007(1):15–15, 2007.
237. Xilinx. Platform studio and the edk. *http://www.xilinx.com/*, Jan. 2011.
238. CoFluent. *http://www.cofluent.com*, Jan. 2011.
239. F. Herrera and E. Villar. A framework for heterogeneous specification and design of electronic embedded systems in SystemC. *ACM Transactions on Design Automation of Electronic Systems*, 12(3):1–31, 2007.
240. F. Herrera, H. Posadas, P. Sanchez, and E. Villar. Systemic embedded software generation from systems. In *Proceedings of the International Conference on Design, Automation and Test in Europe (DATE)*, page 10142, Washington, DC, USA, 2003. IEEE Computer Society.
241. G. Behrmann, A. David, and K.G. Larsen. A tutorial on uppaal. In M. Bernardo and F. Corradini, editors, *4th International School on Formal Methods for the Design of Computer, Communication, and Software Systems*, number 3185 in LNCS, pages 200–236. Springer-Verlag, Berlin, Heidelberg, Sept. 2004.
242. S. Künzli, A. Hamann, R. Ernst, and L. Thiele. Combined approach to system level performance analysis of embedded systems. In *International Conference on Hardware Software Codesign CODES/ISSS*, pages 63–68, Salzburg, Austria, 2007.
243. S. Perathoner, E. Wandeler, L. Thiele, A. Hamann, S. Schliecker, R. Henia, R. Racu, R. Ernst, and M.G. Harbour. Influence of different abstractions on the performance analysis of distributed hard real-time systems. *Design Automation for Embedded Systems*, 13(1):27–49, June 2009.
244. R. Alur and D.L. Dill. A theory of timed automata. *Theoretical Computer Science*, 126: 183–235, 1994.
245. M. Hendriks and M. Verhoef. Timed automata based analysis of embedded system architectures. In *Proceedings of WPDRTS*, pages 8 pp.–, April 2006.
246. ARC International. *http://www.arc.com/*, Jan. 2011.
247. MIPS Technologies Inc., Pro Series Family. *http://www.mips.com/*, Jan. 2011.
248. Target Compiler Technologies. *http://www.retarget.com/*, Jan. 2011.

249. R. Leupers. *Code Optimization Techniques for Embedded Processors – Methods, Algorithms, and Tools*. Kluwer Academic Publishers, Dordrecht, Nov. 2000. ISBN 0-7923-7989-6.
250. T. Kempf, M. Dörper, R. Leupers, G. Ascheid, and H. Meyr (ISS Aachen, DE); T. Kogel and B. Vanthournout (CoWare Inc., BE). A modular simulation framework for spatial and temporal task mapping onto multi-processor soc platforms. In *Proceedings of the International Conference on Design, Automation and Test in Europe (DATE)*, Munich, Germany, March 2005.
251. T. Kempf, E.M. Witte, O. Schliebusch, G. Ascheid, M. Adrat, and M. Antweiler. A concept for waveform description based SDR implementation. In *4th Karlsruhe Workshop on Software Radios (WSR'06)*, Karlsruhe, Germany, March 2006.
252. G. Schirner, A. Gerstlauer, and R. Domer. Abstract, multifaceted modeling of embedded processors for system level design. In *Proceedings of the Asia South Pacific Design Automation Conference (ASPDAC)*, pages 384–389, 2007.
253. A. Bouchhima, I. Bacivarov, W. Youssef, M. Bonaciu, and A.A. Jerraya. Using abstract CPU subsystem simulation model for high level HW/SW architecture exploration. In *Proceedings of the Asia South Pacific Design Automation Conference (ASPDAC)*, pages 969–972, 2005.
254. A. Gerstlauer, H. Yu, and D.D. Gajski. RTOS modeling for system level design. In *Proceedings of Design, Automation and Test in Europe Conference and Exhibition*, pages 130–135, 2003.
255. K. Karuri, M.A. Al Faruque, S. Kraemer, R. Leupers, G. Ascheid, and H. Meyr. Fine-grained application source code profiling for ASIP design. In *42nd Design Automation Conference*, Anaheim, California, USA, June 2005.
256. T. Kempf, E.M. Witte, V. Ramakrishnan, G. Ascheid, M. Adrat, and M. Antweiler. SDR baseband processing portability: A case study. In *SDR'08*, Washington, D.C., USA, Oct. 2008.
257. D.B. West. *Introduction to Graph Theory*. Prentice Hall, Englewood, Cliffs, NJ, 2nd ed., Aug. 2000.
258. T. Kempf, E.M. Witte, V. Ramakrishnan, G. Ascheid, M. Adrat, and M. Antweiler. An SDR implementation concept based on waveform description. *FREQUENZ: Journal of RF-Engineering and Telecommunications, Berlin*, (9-10), 2006.
259. Y.-K. Kwok and I. Ahmad. Static scheduling algorithms for allocating directed task graphs to multiprocessors. *ACM Computer Survey*, 31(4):406–471, 1999.
260. C.D. Locke, E.D. Jensen, and H. Toduda. A time-driven scheduling model for real-time operating systems. In *IEEE Real-Time Systems Symposium*, pages 112–122, 1985.
261. O. Moreira, F. Valente, and M. Bekooij. Scheduling multiple independent hard-real-time jobs on a heterogeneous multiprocessor. In *EMSOFT '07: Proceedings of the 7th ACM and IEEE International Conference on Embedded Software*, pages 57–66, New York, NY, USA, 2007. ACM.
262. J. Leung, L. Kelly, and J.H. Anderson. *Handbook of Scheduling: Algorithms, Models, and Performance Analysis*. CRC Press, Inc., Boca Raton, FL, USA, 2004.
263. A. Papoulis and S.U. Pillai. *Probability, Random Variables and Stochastic Processes*. McGraw-Hill, NY, USA, 4 ed., 2002.
264. The Multicore Association. Multicore Communications API Specification V1.063 (MCAPI), *http://www.multicore-association.org/*, March 2008.
265. Poly Core Software. Poly-Messenger. *www.polycoresoftware.com/*, Dec. 2010.
266. Texas Instruments Inc. Dsp bios kernel. *http://focus.ti.com/*, Jan. 2011.
267. T. Kempf, S. Wallentowitz, G. Ascheid, R. Leupers, and H. Meyr. RWTH Aachen University. A workbench for analytical and simulation based design space exploration of software defined radios. In *VLSI Design Conference 2009*, New Delhi, India, Jan. 2009.
268. D. Piergentili and D. Coupe. Esl methods for optimizing a multi-media phone chip. *EDA Design Line, http://www.edadesignline.com/*, May 2008.
269. T. Kempf, K. Karuri, S. Wallentowitz, G. Ascheid, R. Leupers, and H. Meyr. A SW performance estimation framework for early System-Level-Design using fine-grained instrumentation. In *Proceedings of the International Conference on Design, Automation and Test in Europe (DATE)*, Munich, Germany, March 2006.

270. L. Devroye. *Non-Uniform Random Variate Generation*. Springer Verlag, New York, 1986.
271. C. Lattner and V. Adve. LLVM: A compilation framework for lifelong program analysis and transformation. In *Proceedings of the 2004 International Symposium on Code Generation and Optimization (CGO'04)*, Palo Alto, California, March 2004.
272. A. Bouchhima, P. Gerin, and F. Pétrot. Automatic instrumentation of embedded software for high level hardware/software co-simulation. In *Proceedings of the Asia South Pacific Design Automation Conference (ASPDAC)*, pages 546–551, Piscataway, NJ, USA, 2009. IEEE.
273. A.S. Tanenbaum. *Modern Operating Systems*. Prentice Hall, Upper Saddle River, NJ, 2nd ed. 2001.
274. Xilinx. Microblaze processor reference guide. *http://www.xilinx.com/*, Jan. 2011.
275. J. Tourley. Survey says: Operating systems up for grabs. Technical report, Embedded Systems Design, Embedded.com, May 2005.
276. Linux Kernel. *http://www.kernel.org/*, Jan. 2011.
277. Portable Operating System Interface for uniX (POSIX). IEEE Standard 1003, *http://standards.ieee.org/*, Jan. 2011.
278. Common Object Request Broker Architecture (CORBA), http://www.corba.org/, Jan. 2011.
279. E. Lusk, W. Gropp, and A. Skjellum. Using mpi-portable parallel programming with the message-passing interface. *Science Programme*, 5(3):275–276, 1996.
280. Objective Interface Systems (OIS), Inc., ORBexpress Common Object Request Broker Architecture (CORBA), http://www.ois.com/, Jan. 2011.
281. A. Gill. *Introduction to the Theory of Finite-State Machines*. McGraw-Hill, New York, 1962.
282. T. Kogel, M. Doerper, T. Kempf, A. Wieferink, R. Leupers, and H. Meyr. Virtual architecture mapping: A systemc based methodology for architectural exploration of system-on-chips. In *IJES, Vol. 3, Nr. 3*, pages 150–159, 2008.
283. H. Schildt, American national standards institute, international organization for standardization, international electrotechnical commission, and ISO/IEC JTC 1. *The Annotated ANSI C Standard: American National Standard for Programming Languages C: ANSI/ISO 9899-1990*. 1990.
284. *ISO/IEC 14882:2003: Programming languages: C++*. 2003.
285. Virtual Platform CoWare. *http://www.coware.com/*, Jan. 2011.
286. Zeligsoft. *http://www.zeligsoft.com/*, Jan. 2011.
287. Communications Research Centre Canada (CRC). Scari Software Suite, *http://www.crc.gc.ca/*, Jan. 2011.
288. T. Kempf, E.M. Witte, V. Ramakrishnan, G. Ascheid, M. Adrat, and M. Antweiler. A Workbench for Waveform Description based SDR Implementation. In *2007 Software Defined Radio Technical Conference (SDR'07)*, Denver, USA, Nov. 2007.
289. Extensible Markup Language (XML). *http://www.w3.org/*, Jan. 2011.
290. Texas Instruments Inc. TMS320C55x DSP CPU Reference Guide (Rev. F). User Guide, Feb. 2004.
291. Texas Instruments Inc. TMS320C64x/C64x+ DSP CPU and Instruction Set Reference Guide (Rev. H). User Guide, Oct. 2008.
292. Texas Instrument. DSP Libraries for TMS320C64x and TMS320C55x. *http://www.ti.com/*, Jan. 2011.
293. K. Kennedy and J.R. Allen. *Optimizing Compilers for Modern Architectures: A Dependence-based Approach*. Morgan Kaufmann Publishers Inc., San Francisco, CA, USA, 2002.
294. S.S. Muchnick. *Advanced Compiler Design and Implementation*. Morgan Kaufmann Publishers Inc., San Francisco, CA, USA, 1997.
295. M. Hohenauer, R. Leupers, O. Wahlen, et al. An executable intermediate representation for retargetable compilation and high-level code optimization. In *International Workshop on Systems, Architectures, Modeling, and Simulation (SAMOS)*, 2003.
296. L. Gao, J. Huang, J. Ceng, R. Leupers, G. Ascheid, and H. Meyr. TotalProf: A fast and accurate retargetable source code profiler. In *International Conference on Hardware/Software Codesign and System Synthesis (CODES-ISSS 2009)*, Grenoble, France, 2009.
297. F. Petrot. Automatic timing annotation of native software for mpsoc simulation. In *MP-SoC'08*, June 2008.

298. The MathWorks Inc. MATLAB. *http://www.mathworks.com/*, Jan. 2011.
299. MIL-STD-188-110B Departement of Defense Interface Standard. April 2000.
300. H. Meyr, M. Moeneclaey, and S.A. Fechtel. *Digital Communication Receivers: Synchronization, Channel Estimation and Signal Processing*. Wiley, New York, Feb. 1997.
301. Tensilica Inc. Diamond Standard Processor Core Family Architecture. White Paper, July 2007.
302. A. Viterbi. Error bounds for convolutional codes and an asymptotically optimum decoding algorithm. *IEEE Transactions on Information Theory*, 13(2):260–269, 1967.
303. C. Berrou and A. Glavieux. Near optimum error correcting coding and decoding: Turbo-codes. *IEEE Transactions on Communications*, 44(10):1261–1271, Oct. 1996.
304. Texas Instruments. TMS320C645x DSP Viterbi-Decoder Coprocessor 2 Reference Guide. *http://www.ti.com/litv/pdf/spru972*, April 2006.
305. Texas Instruments. DSP Libraries for TMS320C64x and TMS320C55x. *http://www.ti.com/*, Jan. 2011.
306. Texas Instruments. Reed Solomon Decoder: TMS320C64x Implementation, December 2000.
307. CoWare Task Modeling and Virtual Processing Unit User's Guide. *http://www.coware.com*, Jan. 2011.
308. Texas Instrument. C6flo graphical software development tool. *http://www.ti.com/*, Jan. 2011.

Index

CPSIA information can be obtained at www.ICGtesting.com
225415LV00003B/21/P